Technologies
for basic needs

Technologies
for basic needs

Hans Singer

International Labour Office   Geneva

ISBN 92-2-101774-5 (limp cover)
ISBN 92-2-101773-7 (hard cover)

First published 1977

Printed by Atar S.A., Geneva, Switzerland

FOREWORD

Technology—the skills, knowledge and procedures for providing useful goods and services—has been a principal underlying theme of ILO thinking and activities since the earliest years of the Organisation's existence. The ILO was indeed born out of the process of industrialisation in Europe and North America in the nineteenth century, a process inspired by the continuous development of new technology. And throughout the present century the pace of technological innovation has continued, even accelerated—an apparently independent and ineluctable force. Often, however, the application of the findings of technological research brings in its train not only social benefits but also social costs. For this reason, technological progress has been the direct or indirect stimulus for many of the ILO's longest established activities.

But is the industrialisation process, as it is known in the developed countries, appropriate to the special situation of the developing countries? A great deal of thought has been put into answering this question in recent years. Since the late 1960s, and particularly since the launching of the ILO World Employment Programme in 1969, it has been increasingly recognised that employment-oriented development must not be based only on the direct transfer to the developing countries of the technology found in the industries of the developed world. Moreover, the industrialisation process contributes little to the alleviation of poverty, whether in urban or in rural areas. Generally, the benefits of modern sector growth accrue to the modern sector alone, and in many developing countries industrialisation remains largely irrelevant to the really pressing problems of our times. Hence the need for a redefinition of the place of industrialisation and industrial technology in the development process, and for the careful examination of alternative policies for technology, not only in industry but also in agriculture, construction and rural development.

This book, inspired by the World Employment Conference held in Geneva in June 1976 and by the discussions at that Conference on the agenda item "Technologies for productive employment", suggests new criteria for establishing socially oriented technology policies in the developing economy.

In the Programme of Action adopted by the World Employment Conference a middle way is suggested: "developing countries should arrive at a reasonable balance between labour-intensive and capital-intensive techniques, with a view to achieving the fundamental aim of maximising growth and employment and satisfying basic needs..." (para. 49).[1] Does such a middle course exist? Is it in fact possible to reduce the dependence of developing nations on technologies transferred from the already industrialised world? The affirmative answers to these questions given in this book are based on investigations carried out under the World Employment Programme and elsewhere. These have demonstrated the feasibility of modern, but relatively labour-intensive, technologies in various economic activities, including manufacturing industry, construction and agriculture.

The ILO intends to continue its work on the complex relationships between technological choice, employment and basic needs, by research and by technical co-operation projects to assist the implementation of appropriate technology in practice. Recent developments, in particular the approval by the United States Congress of a "Proposal for a Program on Appropriate Technology" in July 1976, and the World Bank statement in the same month on its activities in this field, have confirmed the coming-of-age of "appropriate technology". I believe that we can reasonably claim that our own efforts in the ILO have made some contribution to the widespread acceptance of a once rather novel concept. I should perhaps add, however, that the ILO has neither the mandate nor the intention to work in isolation on the investigation and promotion of appropriate technology. This is a field in which I attach great importance to close collaboration with other agencies in the United Nations system.

Technologies for basic needs elaborates upon the relevant policy recommendations of the Programme of Action adopted by the World Employment Conference, developing the tentative ideas set out in my report to the Conference.[2] It also sets the scene for further ILO activities in the future. For example, we shall increase our efforts to disseminate information on appropriate technologies for rural development, for small-scale and agro-based industries and for construction; guidelines will be developed to assist managers and management trainers to make technological choices taking into account their employment potential.

Technology should be regarded as the servant of social and economic objectives, and not the master; scientific and technological efforts should be directed towards the improvement of the welfare of the villager, the peasant and the worker in small-scale industry, as well as of modern sector

---

[1] The Declaration of Principles and Programme of Action adopted by the World Employment Conference are reprinted in Appendix A.

[2] ILO: *Employment, growth and basic needs: A one-world problem,* Report of the Director-General, Tripartite World Conference on Employment, Income Distribution and Social Progress and the International Division of Labour (Geneva, 1976), especially Ch. 9, "Technological choice and innovation for developing countries".

managers and employees. Technology, in policy and in practice, should be related to the objective of satisfying basic human needs, the development objective endorsed by the Declaration of Principles of the World Employment Conference. It is this cause, and that Declaration, that the present volume is intended to serve.

Geneva, June 1977

Francis Blanchard,
Director-General,
International Labour Office.

CONTENTS

APPROPRIATE TECHNOLOGY AND BASIC NEEDS

1

SOME CONCEPTUAL ISSUES

The first and obvious determinant of a country's appropriate technology is its factor endowment—that is, the relative proportion in which labour, capital, land, skills and natural resources are available to the economy.[1] For a country which suffers from a relative shortage of capital and a relative abundance of unskilled labour (the typical situation for a developing country), it is possible to choose a technology *in a given sector* or *for a given project* which uses more capital and less unskilled labour than would be justified by the factor proportions available to the economy. However, in that case it follows that some other sector or some other project will have less capital per unit of labour than the average for the economy as a whole.

If a relatively labour-intensive technology can be chosen for all sectors and all projects, this will make for a more unified or integrated economy, with a fair degree of equality of labour productivity in different sectors creating a favourable situation for a reasonable equality of income distribution. If, on the other hand, access to scarce capital or other scarce "resources" (for example, foreign exchange or planning skills) is liberally given on a privileged basis to a certain sector or group of projects, these sectors or groups of projects will form a modern or "formal" sector which will be counter-balanced by a capital-short or "informal" sector. Unless the over-all social structure or the intervention by the government through fiscal and other policies prevents this, such a situation will make for a non-integrated or dualistic structure of the economy, with inherent inequalities of income distribution and a likelihood that those outside the modern sector will fail to satisfy their basic needs.

The technology used in the rural and urban informal sectors is by necessity less capital-intensive and less supported by other inputs (i.e. more labour-intensive) than would in any case be required by the resource endowment of the country. This means that labour productivity is likely to be low,

---

[1] Including capital, skills, etc., available from outside for certain purposes and under certain conditions.

although there may well be groups and subsectors where sufficient technological adjustment to their lack of access to capital has been made, so that living standards may rise above the basic-needs minimum and in fact may be higher than those of some poorer sections within the modern sector.

The technology used in the informal sector will often be nearer to an "appropriate technology" than that used in the modern sector, although in principle both are inappropriate in the sense that the one uses too much capital relative to what the factor endowment would permit while the other, conversely (and consequently), uses too little capital.

In strict logic, it is therefore true that the technology required would be an "intermediate" technology—intermediate between the high capital and related requirements of the modern large-scale sectors and the technology of the small-scale and informal sectors. But there is some question whether an efficient intermediate technology in this sense now exists and whether its creation is justified on cost-benefit or time-factor considerations. Moreover, it is not certain that the application of a uniform intermediate technology throughout the economy is preferable to the existence of areas of "modernity" and high labour productivity, provided that the income distribution can be isolated, to some extent at least, from the effects of uneven technology. On the other hand, where efficient intermediate technology can be shown to exist, or to be capable of being created at reasonable cost, such applications would deserve high priority in development and technology policy.

Thus, one school of thought (prevalent in the 1950s and 60s) held that factor endowment determines appropriate technology, while the actual choice of technology determines factor *use,* the difference between factor endowment and factor use representing unemployment and underemployment. The recommended government action is on factor prices in order to let the factor endowment determine the actual technology.

A more recent view, prompted by actual country and case studies, is that while factor endowment determines appropriate technology, the choice of technology determines the evenness or unevenness of technology in different sectors of the economy and hence determines income distribution. In turn, income distribution determines the actual technology via the product mix. The recommended government action in this case is either on income distribution as an indirect determinant of the output mix or directly on the output mix by fiscal or other measures. Export orientation or internal self-reliance could influence the output mix and scale in a labour-intensive direction, while simple import substitution, with the policies associated with this strategy, would have the opposite effect. *True* import substitution, i.e. the replacement of inappropriate imported products by appropriate local products, using local resources and labour, would remain a keystone of the new strategy.

APPROPRIATE TECHNOLOGY AND BASIC-NEEDS STRATEGY

It is now becoming increasingly clear that conventional development strategies which emphasise the growth of gross national product *per se,* without at the same time inquiring into the pattern of growth which determines its fruits, do not, in most developing countries, alleviate mass poverty and unemployment. What has happened in many cases is that through conventional growth strategies the fruits of growth have been concentrated in the hands of a small privileged minority and have not reached the bulk of the population. The reasons for this state of affairs are, however, complicated and beyond the scope of the present volume. One of the factors which seems to have contributed to the perpetuation of poverty is that rapid growth has occurred in the small modern sector of the economy using most advanced imported technology. This growth has not spilled over into the rural traditional and urban informal sectors. In fact, quite often growth in the modern sector has occurred at the expense of these sectors. Technological progress in the former often does not lead to the raising of technological levels in the latter.

If past patterns of development have not yielded the desired results, there is clearly a need for current development strategies to be reoriented towards the elimination of poverty and unemployment and the fulfilment of basic needs. These three elements are all inter-related. Both unemployment and underemployment prevent the majority of the population in developing countries from having access to minimum personal consumption needs such as adequate food and shelter, and to minimum social services such as water, education, sanitation, medical facilities and transport. Thus the technology required for a basic-needs strategy in a developing country must concentrate more than in the past on meeting the requirements of the small farmer, small-scale rural industry and the informal sector producer. Such a strategy calls for, and is in turn supported by, a special kind of appropriate technology: a technology which differs from that developed in the industralised countries by the industrialised countries and for the industrialised countries even more than the difference in factor proportions would require. This is so because under a basic-needs strategy technology must bear the double burden of adapting existing or imported new technology to the general situation of the developing country *and* of underpinning the redistribution of incomes which goes with a basic-needs strategy. For this reason it might be called a "doubly appropriate" technology. It can safely be assumed that a "doubly appropriate" technology must contain a greater element of technological innovation (although possibly based on pre-existing knowledge not currently selected for use or development by the industrialised countries) than a "simple appropriate" technology, which can more often use the instruments of selective choice and adaptation applied to existing technologies, usually developed in the industrialised countries.

If the rural and the urban poor are the target groups in a basic-needs strategy, for whose especial benefit scientific and technological knowledge is to be put to use, clearly the technologies which are imported from abroad may not always be appropriate to their requirements. If the technological levels in the rural and urban informal sectors are to be raised (as is, in fact, called for in the Programme of Action adopted by the ILO World Employment Conference), it is essential to adopt those technologies to which the small farmers, artisans and other small producers have easy access with their limited cash resources. This is not to suggest that modern technological know-how is not relevant or important, but only that a selective approach to the adaptation and adoption of known methods is needed. In fact, the experience of countries which have tried to implement a basic-needs strategy (e.g. China, Cuba, Tanzania) suggests that the improvement of simple village technologies is the only feasible approach to the gradual modernisation of the rural economy. The experience of Tanzania, which was examined in detail in an ILO/UNDP technical assistance project on appropriate technology, shows that in a subsistence economy the initial cash outlays required for imported equipment are far in excess of what the poor farmers can afford. It is therefore imperative to utilise local resources and skills for the design and development of technologies that are more productive than the traditional ones and yet are within the reach of farmers and other poverty groups.

The implications of a change in development aims away from GNP growth and towards the new objectives of a basic-needs strategy have been clearly spelt out as follows:

The crucial task for every government is to develop technical expertise and to control consumer aspirations.... Outside the "least developed" countries, the universal need is for sufficient technical and administrative capacity to screen the importation of production and consumption techniques, and monitor those which have been allowed into the country.

It follows that a planning office should no longer be primarily concerned with growth targets or with projects, but with creating the cadres capable of evolving and implementing a development strategy expressed in targets for key resources and styles of consumption. Its second priority would be building up the information needed to back such a strategy, especially in negotiations with foreign firms and governments— information on the range of technologies available internationally, on the cost of different ways of obtaining them, etc. [1]

This statement clearly confirms the enhanced importance of technological policy in national planning. In the light of earlier discussions, one would add the more positive functions of adapting (rather than "monitoring") the imported technologies and developing an indigenous technology as one of the cornerstones of national development. The purpose of technological policy is to reconcile the objectives of spreading employment and income with the fast development of key sectors.

The enhanced importance of technological capability as a development objective *per se* should not obscure the fact that basically it is the development

---

[1] Dudley Seers: "A new look at the three world classification", in *IDS Bulletin* (Brighton, University of Sussex), Vol. 7, 1976, No. 4.

planners who set the concrete technological tasks and allocate priorities: emphasis on rural or urban development, on a high rate of investment or promotion of consumption, high-level consumption or basic needs, priority for the development of specific regions of the country, export promotion or import substitution, and so forth. Each planning decision of this sort, whether explicit or implicit in the projects proposed, will have a consequential impact on technological policy. The same is true of major macro-economic policies: fiscal, monetary and foreign exchange policies, wage policies, land reform policies, policies respecting foreign investment—all these influence the adjustment of technological policy.

ELEMENTS OF AN APPROPRIATE TECHNOLOGICAL POLICY [1]

The first decision to be made is the choice of technology for a given project or range of activities. Choice involves a range of alternatives. How wide is the range of alternatives, or the "menu", from which technology can be chosen? The hypothetical range, or number of items on the "menu", is clearly very large. In so far as technology is embodied in equipment packages, it would be scientifically and technically possible to produce an immense range of such equipment packages involving different degrees of capital intensity. In the past the industrialised countries have done so in the course of their industrial history, moving steadily to equipment and technologies which became steadily both more efficient and more capital-intensive (in line with their changing factor proportions). It is this combination of increased efficiency, on the one hand, and increased capital intensity, on the other, which has led to the popular association of these two attributes—a belief that increased efficiency and increased capital intensity are identical. Although there is a clear *historical* or secular association of the two attributes, there is no strict *logical* association. In terms of science and technology, there is no reason to assume that greater efficiency must be labour-saving rather than capital-saving.

But not all the hypothetical wide-ranging possibilities are in fact available. Only a portion of the range is actually investigated, only an even narrower range is actually brought to the testing or blueprint or prototype stage, and the range of equipment actually produced is again a narrower selection from the investigated range. Moreover, the historical range is now partly or largely submerged: only a part of the discontinued historical technology can be revised or reconstructed on the basis of available blueprints or other information, and this at varying cost. Technical progress destroys as well as creates knowledge. A developing country with its own capital goods industry and/or

---

[1] This section observes the terminology and framework established in Bruce F. Johnston and Peter Kilby: *Agriculture and structural transformation: Economic strategies in late-developing countries* (New York, London and Toronto, Oxford University Press, 1975), especially pp. 87-92 and 105-114.

strong national technological capacity obviously has a much wider range of possibilities actually available to it than one without these two assets.

Since the world is fairly clearly divided into technology-initiating and technology-borrowing countries, the technology available to the latter depends to a greater or lesser extent on the technologies which are available and activated in the technology-initiating countries. Those activated in earlier periods will be embodied in older equipment (in so far as they can still be produced or are available from stock) or in second-hand equipment. Both are important means of extending the range of technology that is likely to be useful to developing countries, but both also have their specific problems.[1] The range of technology that is actually available at any given moment forms the "technology shelf"—that is, the items on the "menu" which the kitchen will actually supply on order.

Analysts have differed in their estimates of the breadth of the range of available technologies, although in general they have been in agreement over the variation of capital/labour ratios in different industries and sectors. A more neo-classical model is based on the assumption of a wide range of available technologies, and failures of choice are then traced back to internal factors in the technology-borrowing developing country: for instance, failure to adjust factor prices or to use shadow factor prices, unequal income distribution, overvalued exchange rates, biases in training, lack of skill in negotiations, and so on. A more institutional or "deterministic" model will emphasise the lack of choice, the narrowness of the range, with which developing countries are confronted, and this is then traced back to conditions in the technology-initiating countries, or in the world economy as a whole, creating one-sided technological dependence. Obviously there are elements of reality in both these views. In any case, the term "available" is itself ambiguous and open to differing interpretations: indeed, the two schools of thought noted above tend to interpret this term in different ways.

As has been argued above, the use of capital-intensive technology in capital-short economies is bound to lead to a dualistic pattern of development that is not easily reconciled with a basic-needs strategy. Even within production sectors (e.g. industry), what has been described as a "discontinuous bimodal distribution of factor intensities"[2] may result: a cluster of large-scale capital-intensive plants (perhaps dominated by foreign investors) and another cluster of small-scale labour-intensive producers, verging into what the ILO comprehensive employment strategy mission to Kenya described as the "informal sector".[3] Thus, by putting more emphasis on the more labour-intensive small-scale sector, yet another element is introduced into technological choice: that of scale of production.

---

[1] See Chapter 4 below for a more detailed discussion of second-hand machinery.

[2] Johnston and Kilby, op. cit., p. 91.

[3] ILO: *Employment, incomes and equality: A strategy for increasing productive employment in Kenya* (Geneva, 1972), Ch. 13 and technical paper 22.

One way in which the range of available technologies can be extended is by "unpackaging" the production process. Even where the primary or core process may be constrained by the need to choose a capital-intensive technology from a narrow range of available technologies, this need not apply to the ancillary or peripheral parts of the production process. The moving of materials to and from the core process would be a typical illustration of such an ancillary activity within the production process which lends itself to variations from the standard technology of industrial countries that are both more flexible and more labour-intensive. This and other methods of adaptation are discussed in Chapter 4 below.

Another method of adaptation is through product variation. This covers a broad spectrum from minor qualitative differences (e.g. the colour of the crust of bread baked in a simpler oven) to essentially different products (e.g. fashion leather shoes versus rubber sandals made from discarded tyres). This method of achieving a broader range of technological choice has been emphasised and described as one of complementing "appropriate technology" by "appropriate products", and it is particularly closely associated with a basic-needs strategy. Generally speaking, the simplification of products will make possible the use of more labour-intensive technology while at the same time broadening markets and benefiting larger numbers of people. Where income inequalities are great, the commercial incentive for research on appropriate technologies is reduced. A reduction in income inequality and the development of appropriate products can therefore be made mutually reinforcing.

Another method, that of process simplification, is often closely linked with the development of new technologies, discussed in Chapter 5 below, since the new machine is simultaneously both simplified and different. That is true of both the examples given by Johnston and Kilby: the Nigerian dough-brake used in breadmaking, and the diesel engine used in India and Pakistan for lifting water out of wells and canals. The former is constructed from reconditioned scrap and used material in place of expensive inputs. The diesel engine uses cheaper fuel and has a longer life and fewer breakdowns; it goes back to an older technology, and also substitutes scrap for pig iron. [1]

A final method of broadening the range of technologies available is by "labour-addition". This includes the combination of a given core of equipment with more labour to reduce breakdowns, use cheaper materials more prone to breakdowns, run equipment faster, etc., and to make possible the greater use of equipment through multiple-shift working.

If the various methods briefly surveyed in this section—local capital goods industry, unpackaging the production process, varying the scale of production, the nature of the product and rate of capacity utilisation, process simplification, use of older and second-hand equipment—are all considered "available" for the variation of technology, one would be bound to move

---

[1] Johnston and Kilby, op. cit., pp. 111-112.

closer to the neo-classical model in assuming a wide choice of technology. However, these extensions of the narrower concept of "available range" require a certain degree of managerial, administrative and technological maturity, and also a framework for an effective over-all development policy, such as a basic-needs strategy.

APPROPRIATE TECHNOLOGY, TECHNOLOGICAL DEPENDENCE AND THE TECHNOLOGICAL GAP

The idea of "appropriate technology" sometimes meets with the criticism that it carries a thinly veiled "neo-colonial" implication—an implication that the existing, modern, efficient technology is not right for the developing countries, which should be satisfied with something inferior, second-best, less efficient. This is a serious reaction which merits careful study.

To begin with, one must recognise that there is an element of truth in this objection. If the situation is such that all technological power is concentrated in the rich countries, and that therefore the only efficient technology developed and existing is the capital-intensive and sophisticated technology, then it is quite true that the use of this technology is inevitable, even though it may not be ideally suited to the needs and requirements of developing countries. Its superior efficiency would outweigh its inappropriateness. In that case, it is very proper that all efforts be concentrated on achieving a transfer of this modern technology under the best possible conditions, free of undue restrictions and involving the least possible drain of other resources. That certainly is the situation in certain sectors (for example, oil refineries or atomic energy installations) where safety and precision and other related product requirements are supreme. But those who advocate appropriate technology, while quite ready to admit that such areas exist, would argue: *(a)* that this is not the situation in all or most sectors of the economy; and *(b)* that even where it does exist it is only a second-best solution for the developing countries. The best solution would be to develop a technology which is at once efficient and modern and yet better geared to the resources and requirements of developing countries.

Those who are suspicious of the idea of an appropriate technology sometimes also argue as follows: "The industrialised countries are rich; they use a modern, sophisticated, capital-intensive technology; *ergo,* they must be rich because they use this particular technology." If the idea is put in this form, the error in the reasoning will spring to the eye. The industrialised countries are not rich because they use a capital-intensive technology. On the contrary, one could argue that if in the early stages of their industrial revolutions the industrialised countries had tried to use the technology which they rightly use today they would never have become as rich as they are now (although this is an hypothetical argument that cannot be proved one way or the other). The industrialised countries are not rich because they use the sophisticated, capital-intensive technology; the line of causation is the other way round. They use the capital-intensive technology—and rightly so—because they are rich.

The suspicion of appropriate technology is increased when the proposal takes the form of suggesting an intermediate technology. In fact, the appropriate technology in a number of important respects, such as capital intensity, will be intermediate between the traditional technology now prevailing over the activities of most people in developing countries and the modern, capital-intensive technology widely prevailing in the industrialised countries. But the idea of "intermediate technology" can be misinterpreted as suggesting a technology which is intermediate in efficiency between the low efficiency of present traditional technologies and the high efficiency of modern technology. This is not in the mind of those who use the term "intermediate technology"; they rightly emphasise that the intermediate technology they advocate is in fact the most efficient for the circumstances of most developing countries. All the same, it is preferable to speak of "appropriate technology" rather than "intermediate technology".

The criticisms of appropriate technology may be further allayed by emphasising that it is a transitional policy and not a permanent policy. As developing countries succeed in achieving development, their factor proportions will become more similar to those of the industrialised countries, and the difference in the technologies appropriate for the two groups of countries will diminish and perhaps finally disappear. One can say that the purpose of those who emphasise that the technology appropriate for developing countries is different from the technology appropriate for rich industrialised countries is precisely to make their own statement redundant. As development is successfully achieved by means of appropriate technology, the need for a different appropriate technology for developing countries will gradually disappear.

Again, the critics of the concept sometimes argue that it in effect establishes two different standards and will therefore create and perpetuate a "technological gap". This is a misconception. The gap exists in the fact that some countries are poor while other countries are rich. The task is to reduce or eliminate *this* gap—the *economic* gap. Different technologies will serve to reduce the economic gap and hence, ultimately, to eliminate the need for different technologies, as just argued. If we misdefine the problem by declaring that the gap is a *technological* gap, and then try (disregarding the economic gap) to apply exactly the same technology to the two groups of countries, the real economic gap will widen further instead of narrowing. Thus, to say that the application of the concept of appropriate technology perpetuates the gaps between rich and poor countries is a travesty of the true position.

The need for an appropriate technology different from that of industrialised countries may not be universal among the developing countries. For some of the OPEC countries, for example, the abundance of foreign exchange reserves is such that the acquisition of modern technology and the introduction of capital-intensive methods of production would certainly be considered a reasonable strategy. Where small populations are combined with very large liquid capital assets the factor proportions are in fact highly similar to those of industrialised countries.

Finally, the discussion of appropriate technology—and the concept that the appropriate technology for developing countries is different from that for industrialised countries—should by no means obscure the desirability and necessity of giving the developing countries the freest possible access to the latest, highly capital-intensive technology, wherever such a technology is inevitable or appropriate because of superior efficiency, or wherever developing countries decide to adopt modern technology. Where this is the case we do not want to see resources drained from poor countries to rich countries because of excessive transfer costs or unreasonable restrictions. This statement is subject to two provisos: *(a)* the use by developing countries of capital-intensive technologies should be limited to those cases where it is necessary and preferable; and *(b)* the lower cost of such basically inappropriate technologies should not lead to their use at the expense of efforts to develop a more appropriate national technology.

The concept of technological dependence versus technological self-reliance is somewhat ambiguous. One aspect of technological dependence is an undue weakness in negotiating with those who own or control modern technology, so that access to modern technology is impeded or becomes too expensive. The other aspect of technological dependence is that it implies the acceptance and use of a technology that is not the most appropriate for the conditions in a given developing country. Both these aspects of technological dependence are valid and important, and will be dealt with in the chapters which follow.

There can be no dispute with the formulation adopted by the World Employment Conference as part of its Programme of Action, i.e. that of a "reasonable balance between labour-intensive and capital-intensive techniques".[1] This is rightly associated with a corresponding balance of goals, such as maximum growth, maximum employment and the satisfaction of basic needs. Where developing countries wish to adopt advanced techniques as an objective *per se,* this would then also be a factor in determining the balance of "equilibrium between the various types of technologies".[2]

---

[1] Para. 49.

[2] ibid.

DETERMINANTS OF A COUNTRY'S TECHNOLOGY MIX

2

Having discussed some of the conceptual problems concerning the choice of technology and its relation to a basic-needs strategy, we now turn to some of the determinants of a country's technology mix. This will serve as a basis for a more systematic discussion of problems and policy conclusions. The major determinants may be listed as follows:[1]

(a) existing technologies: the "technology shelf", from which technologies are selected, transferred and disseminated;

(b) a country's ability to adapt existing technology to its own special or changing conditions;

(c) a country's capacity to create national or indigenous technology suitable for and specifically geared to a country's objectives and circumstances.

The above three factors represent what one might call the "supply" side of technology. There are also certain factors on the "demand" side, including:

(d) the state of factor prices and other incentives facing decision makers;

(e) income distribution, which determines the effective market demand for various products and sectors;

(f) the nature and situation of those who make decisions about technology to be used in concrete projects.

The six determinants identified above provide a convenient classification on the basis of which some of the more general problems of technological choice may be discussed. However, the results of empirical research show that whenever a choice of technology is made, numerous technical, practical and organisational problems enter which cannot be neatly caught by logical classifications. So far, this empirical research is only beginning. Yet it has already unearthed a multitude of practical factors relating to quality of pro-

---

[1] The concept of technology mix and the partitioning of determinants into supply and demand factors follows P. Kilby: "Appropriate technology at the national level: A survey", in ILO: *Tripartite World Conference on Employment, Income Distribution and Social Progress and the International Division of Labour, Geneva, June 1976: Background papers*, 2 vols. (Geneva, 1976), Vol. I: *Basic needs and national employment strategies*.

duct, scale of production, motivations of decision makers, and so on. While the results so far obtained are embodied as far as possible in the subsequent text, they indicate the dangers and difficulties of generalisation in this field at the present stage.

EXISTING TECHNOLOGY: THE "TECHNOLOGY SHELF"

A developing country that is weak in indigenous capacity to create technology, as a result of previous colonial dependence, poverty, small size or a lack of the necessary skills and experience, will in the short run have to rely mainly on existing technology. The immediate problem here is that most existing technology was created by the group of richer, industrialised countries with a long history of and capacity for creating new technologies.

Many problems arise for the low-income countries as a result of this imbalance in technological capacity under the present world economic system—the "old international economic order". It is difficult to quantify the imbalance with any degree of precision since no satisfactory measurements of technological knowledge (nor of technological innovation) yet exist; but, using the amounts spent in the recent past on creating technological knowledge (so-called R and D expenditures) or the availability of trained scientists, technologists and other high-level skilled personnel as a guide, we may conclude that perhaps as much as 98 per cent of the world's modern technological capacity is concentrated in the industrialised countries (which include less than one-third of the world's population), leaving only a minute proportion in the developing countries (with over double the population). On a per head basis, this amounts to a disproportion of about 1:100. While this figure may overestimate the true disproportion since it excludes the technical know-how that is not created through R and D expenditures, there is no doubt but that a huge disproportion does exist. The reduction of this disproportion is one of the chief objectives of the Declaration on the Establishment of a New International Economic Order adopted at the Sixth Special Session of the United Nations General Assembly in 1974 and the related resolution on development and international economic co-operation adopted at the Seventh Special Session in 1975.[1]

It could be objected that this disproportion does not matter. Why should the developing countries not use the technology available even though it was developed in richer and more industrialised countries? Is it not an advantage for them to be able to use existing technology developed at the expense of other countries rather than to have to develop their own? Is this not the historical advantage enjoyed by latecomers in economic development—that they are able to catch up and often overtake the pioneers? These are legitimate questions and they contain an important element of truth, both

---

[1] Resolutions 3201 (S-VI) of 1 May 1974 and 3362 (S-VII) of 16 Sep. 1975.

analytical and historical. It is true that the technological advances made in Great Britain at the start of the industrial revolution rapidly spread to France, Germany, Russia, the United States and other countries: the existence of a technology shelf created by Great Britain was a key factor in their rapid economic development. This was made possible because the resource endowment of these "second wave" countries was sufficiently similar to that of the pioneers and because their technical capacity was sufficient (or, as in the case of Japan, purposefully created) to enable them to absorb and adapt the new technologies and become autonomous centres of technological capacity.

Even today it is true that selection from the existing technology shelf is a prime source of progress and development. Many modern technological processes give products that are essential to low-income countries, such as fertilisers, by methods which may be capital-intensive but which, if used to raise the productivity of small farmers, will in fact help to create widespread indirect employment through fertiliser *use*. In other cases technical progress is capital-saving, skill-saving or resource-saving instead of, or as well as, labour-saving; in such cases the modern technology is clearly superior for developing countries as regards creating employment and reducing poverty as well as increasing the gross national product. The same is true where the modern technology gives a superior product without corresponding increases in cost. It should by no means be assumed that all modern technology is inappropriate. Indeed, some of the obstacles to the transfer and proper flow of existing technology are rightly considered as problems to be resolved for the sake of accelerated development.

Why, then, are there problems in advising developing countries simply to rely on the industrialised countries to provide effective technologies from which they can benefit? It is important to set out the reasons as a basis for policy formulation.

1. The technology developed in the high-income countries is, naturally enough, a technology developed not only by the richer countries and in the richer countries but also for the richer countries. In other words, it tries to solve the problems of the richer countries by methods which are appropriate to *their* situation: relative abundance of capital, relative abundance of high-level skills, relative shortage of a number of natural resources, relative shortage of unskilled labour and absence of an agricultural surplus population. It is therefore understandable that the technology developed in the richer countries should tend to be capital-intensive, skill-intensive, directed towards the development and use of synthetic substitutes for natural resources and, above all, labour-saving. Equally naturally, it is directed towards the development and production of high-income products.

The developing countries, however, have different problems and a different resource situation. Their immediate problem is to increase productive employment and the production of basic-needs goods within the constraints of severe shortages of capital, certain types of skill and, usually, foreign exchange also. Under the present world system there *is* an appropriate technology, and a dynamic capacity to develop improved technologies, for the minority of mankind living in the rich and industrialised countries. But there is no similar technology and capacity for the majority of mankind living in the developing countries. This gap is one of the main reasons for the lack of productive employment, for unequal income distribution and poverty, and for the low level (and often also slow growth) of gross national product. Although different analyses and different schools of thought may emphasise different aspects and propose different ways of dealing with this gap, they would all be agreed on its existence and on the urgency of reducing it in the future.

2. Countries which have a low national or indigenous capacity to create efficient appropriate technology will also experience great difficulties in making appropriate choices from the existing technology shelf. The selection, transfer and dissemination of technologies "off the shelf" is, in itself, an act which requires a national or indigenous technological capacity if it is to be successfully accomplished. To negotiate with machinery salesmen and select the most economical and advantageous equipment, to examine tenders for development projects, to license or approve project proposals using a given technology, to negotiate effectively with foreign investors and multinational enterprises on proposed projects, or to decide on the use of patents—these are just a few of the acts involved in selecting from existing processes that require a high degree of national technological capacity.

Where such capacity is lacking,[1] the governments and nationals of developing countries can no doubt be advised in these matters by ad hoc expatriate advisers recruited by themselves or commissioned to assist them by international organisations. In negotiating aid projects under bilateral or international auspices, the aid donors may no doubt have a general duty and interest in not negating the objective of their aid by persuading or pressuring the aid recipients to adopt inappropriate technology.[2] But the fact remains that the selection and proper use of technology from the existing shelf requires a technological capacity of the same general order as the act of creating new technology. Somewhat overstating the argument, one might say that a country which has the capacity to create its own appropriate technology has much less need to do so than a country without that capacity, because the former country can make better and more efficient use of existing technology. This is one of the paradoxes of the technological gap from which the developing countries suffer.

---

[1] It would be wrong to assume that this lack is general among developing countries. Countries such as Argentina, Brazil, India, the Republic of Korea and Mexico—to name but a few—have considerable capacity of this kind. Other countries (for example, the Andean Pact countries in Latin America) are rapidly acquiring it.

[2] But see Chapter 3 below for some of the problems arising from procurement tying and other forms of tied aid.

3. The unequal global distribution of technological capacity also creates an obvious advantage for those possessing it. Hence, the technological imbalance in the world creates the danger not only that the transfer of technology to developing countries may be inappropriate for the solution of their problems, but also that it may take place on unequal or unfair terms which place an unduly heavy burden on their resources. This was emphasised in the resolution on development and international economic co-operation adopted at the Seventh Special Session of the United Nations General Assembly.[1] This programme called for the evolution of an international code of conduct for the transfer of technology corresponding to the special needs of developing countries, and this question was a major item on the agenda of the Fourth Session of the United Nations Conference on Trade and Development (UNCTAD). A number of instruments used in the transfer of existing technology from industrialised to developing countries, particularly international patents, would equally be in need of revision to make existing technology more accessible to and cheaper for developing countries.

An estimate of the direct or visible foreign exchange cost of technology payments *per se* (payments for patents, licences, know-how, trademarks, management and other technical services) by developing countries indicated that this may have reached US$1,500 million by the end of the 1960s.[2] The indirect payments embodied in the high cost of imported equipment, over-priced intermediate goods and transfer pricing are certainly even higher than the direct payments. Since 1968 the rate of growth of the direct payments alone is estimated at 20 per cent per annum, and this absorbs a rapidly rising proportion of total export proceeds (or adds a rapidly rising proportion to import costs). Together with normal remitted profits these payments go a long way towards offsetting new resource transfers to developing countries, leaving them with a rising debt burden—even if the higher prices paid for oil and other essential imports are disregarded.

4. These monopolistic advantages and the resulting high cost of technology to less developed countries can be reduced by more effective "shopping around" among the various sellers and sources of a desired technology. However, shopping around itself requires considerable resources; it also calls for technological capacity of a high order if the relative advantages and disadvantages of technology available from different sources are to be accurately compared. Moreover, the high price of technology may often be concealed as part of a package, which makes it very difficult to assess the best source of technology *per se*. It has been shown[3] that in certain cases the mechanism of transfer pricing—that is, declaring high values for needed

---

[1] Resolution 3362 (S-VII) of 16 Sep. 1975.

[2] UNCTAD: *Major issues arising from the transfer of technology to developing countries* (New York, United Nations, 1975; Sales No.: E.75.II.D.2).

[3] Especially by C. Vaitsos in various studies of the pharmaceutical and other industries in the Andean Pact area.

imports and/or low values for exports—may constitute a very high multiple of open or visible payments for technology and may represent a heavy burden for developing countries.

5. While the technology itself is biased in the direction of the factor proportions and product attributes appropriate to the industrialised countries, there is the additional factor that the sources of information about what is available and advice on what is suitable are equally biased. It has been pointed out that, for example, equipment salesmen sometimes tend to offer to developing countries the most expensive machinery, which in some cases may be even more labour-saving than equipment currently in use in typical plants in the industrialised countries themselves.[1] Nor is the advice which developing countries can obtain from consulting engineers or other advisers at their disposal likely to be free of such biases. This establishes a clear case for technology-importing countries to endeavour to develop their own consulting and designing—but this again is one aspect of national technological capacity.

The main channels of communication relating to available technologies and of recommendations for the adoption of a specific technology are multinational enterprises, machinery salesmen and consultancy firms from advanced countries, engineers trained in industrialised countries or through textbooks and instruction based on the experience of industrialised countries, and so on. *None* of these has a clear interest in exploring more appropriate labour-intensive or smaller-scale technologies and/or the possibilities of saving capital through the more intensive use of existing equipment or the use of older and possibly second-hand equipment. On the contrary, there may be a clear interest in recommending the most expensive equipment. As with any broad statement, there are obvious complications in real life: for instance, multinational enterprises may well be interested, in specific cases, in using second-hand equipment from their plants in industrialised countries for production in developing countries. But this will be the exception rather than the rule.

Empirical research has shown that the existing technology shelf is in fact wider than has sometimes been thought. Frequently a range of technologies with different labour intensities, scale factors, product quality and other characteristics of interest to low-income countries may exist which is broader than the narrow band of choices of which the decision makers in developing countries are aware. Moreover, it has also been shown that the different degrees of labour intensity of existing technology often make little difference to profitability or the cost of production.[2] Thus, selections from the existing technology shelf can be improved from a basic-needs viewpoint by wider

---

[1] United Nations, Department of Economic and Social Affairs: *Appropriate technology and research for industrial development: Report of the Advisory Committee on the Application of Science and Technology to Development on two aspects of industrial growth* (New York, 1972; Sales No.: E.72.II.A.3), para. 23.

[2] See James Pickett et al.: "The choice of technology, economic efficiency and employment in developing countries", in *World Development* (Oxford), Mar. 1974.

information regarding existing possibilities, and improved methods of decision making.

In this connection, the existence of older types of equipment and technology, which may be obsolete in the industrialised countries but which may in fact be of great benefit to low-income countries in the light of their different present problems and circumstances, should certainly be investigated. It may also be possible to find such beneficial technologies in the history and among the older blueprints of the industrialised countries.[1]

This must not be confused with the proposition that the technology which would and should emerge as a result of the strengthening by the developing countries of their own national technological capacity is necessarily similar to the now obsolete technologies formerly used in industrialised countries. On the contrary, one would conclude both from economic analysis and from empirical evidence that normally appropriate technology would be different not only from the current technology of industrialised countries but also from any technology used in their previous technological history.

Development decisions have to be made every day. Thus, technologies often *have* to be taken from the existing shelf, even when it is realised that they may be second-best choices, since the technology that is really the most suitable does not exist and would have to be indigenously created. Even if the demands of training are disregarded, this would mean a delay of years owing to the need to develop prototype machines, test them and introduce them into the production process, tool up, and so forth. Thus the only possible means of seeing results within a reasonable time would be to adopt existing technologies. Of course, this makes it doubly important that the choice of existing technology is carefully and purposefully made, and based upon the best possible information and advice.

However, if the creation of national technological capacity is indefinitely postponed, the existing technology shelf will tend to become increasingly remote from and inappropriate for the needs of developing countries, given the trend of technical progress in the industrialised countries. Moreover, the resources required for the proper selection and adaptation of existing technology at any given time will also increase and will compete with the resources needed for creating new technology. Here there are obvious reasons for a technological policy on the part of developing countries in which the use of existing technology and the creation of national technological capacity must go hand in hand. All the same, the time factor highlights the need for measures to make existing technology more accessible and cheaper to developing countries. These measures would include, for instance, improvements in the international patent system and the establishment of technology data banks and reference services.

---

[1] Clear evidence of such a situation is given in R. M. Bell et al.: *Industrial technology and employment opportunity: A study of technical alternatives for can manufacture in developing countries* (unpublished).

CAPACITY TO ADAPT EXISTING TECHNOLOGY

The boundary between the "adaptation", on the one hand, and the "selection" or "use" of existing technology, on the other hand, is not very clear. Nor is the boundary between the "adaptation" of existing technology and the "creation" of new technology. Nevertheless, in spite of boundary disputes and even despite the lack of a precise definition, it is useful to realise that there is a middle ground between the use of existing technology and the creation of indigenous new technology. The properly adapted use of existing technology can be a satisfactory compromise, combining the ready availability of existing technology with its adjustment to local needs and conditions. In the case of imported capital-intensive technology, adaptation may consist of adjustment to a smaller scale of production, adjustment of the quality or nature of the product, adjustment of ancillary processes (such as packing, transport within the plant) to the lower cost of labour, use of local materials, and so on. Generally speaking, adaptation will make the existing technology more appropriate to the factor endowment of the developing country.

In agriculture it is normally necessary to adapt existing technologies to the specific climatic and soil conditions, or the specific nature of the local product and the specific modes of agricultural production. In the case of large-scale industry the need for adaptation is less obvious; moreover, there is a danger that the disadvantages of a lack of adaptation may be concealed, and the pressures to adapt may be reduced by adapting instead the factors making for profitability (by granting tariff protection or various forms of fiscal subsidies and concessions, permitting high prices to be charged, lowering the cost of capital and imported inputs through overvaluation of domestic currency, and so on). Such arrangements reducing the pressure for adjustment will at the same time distort the choices from the existing technology shelf and discourage technological innnovation.

As already noted, in the adaptation of existing technology special importance attaches to the upgrading and improvement of existing local technologies in the small-scale and informal sectors, and also in such sectors as services, education and health. To build on existing technology has obvious advantages over the creation of technology *de novo*. It will generally be an easier task; it will be more naturally production-oriented, and the danger of an "academic" remoteness from the production process will be less; and the time lag before production actually starts will also be less. The improvement of existing technologies can also be a stepping-stone towards building up a capacity for innovation, and provide a useful agenda for needed innovations. The spread (often very wide) in the efficiency of technologies actually applied within given sectors provides scope for spreading existing "best practices" more widely, through the adaptation of the practices of the less efficient units. Such differences in efficiency between and within sectors can sometimes be dealt with through subcontracting relationships where the superior technological capacity of the parent firm is enlisted in upgrading the efficiency of the less efficient, usually small-scale, subcontractors.

CAPACITY TO CREATE NEW TECHNOLOGY

There are numerous arguments for reducing the technological imbalance in the world today by increasing the share of the developing countries in technological innovation. This has been recognised as a world objective, and was embodied by the United Nations General Assembly in the programme for a new international economic order which puts the "establishment, strengthening and development of the scientific and technological infrastructure of developing countries" first among the objectives in the section dealing with science and technology.[1]

One must, however, make the qualification that it is not true that all technology created in less developed countries will necessarily and automatically be more appropriate than technology originating in industrialised countries. The dominance of the industrialised countries' technology and the links between scientists and technologists in the less developed countries and the well established and dominant scientific and technological traditions of their colleagues in industrialised countries (which are also reflected in their own training and orientation) are strong. They constitute a constant temptation for those in the developing countries to accept the problems and approaches of the industrialised countries as guides even for the scientific and technological efforts of the developing countries themselves. The smaller the R and D machinery of developing countries in relation to that of the industrialised countries, and the more isolated and scattered the scientists and technologists in the developing countries, the greater is this risk of an "internal brain drain", as it has been described.[2]

This is, of course, in addition to the visible or "external" brain drain [3] which constantly hampers the ability of low-income countries to create their own national technological capacity.

Nor is it true that the creation of a new technology must necessarily be located within the developing countries themselves. In fact, the United Nations World Plan of Action makes specific proposals for deliberately creating new technologies appropriate for developing countries in the industrialised countries, by diverting a specified percentage of their R and D expenditures to this purpose.[4] However, such a development, desirable as it is,

---

[1] Resolution 3362 (S-VII) of 16 Sep. 1975.

[2] United Nations, Department of Economic and Social Affairs: *Science and technology for development: Proposals for the Second United Nations Development Decade* (New York, 1970; Sales No.: E.70.I.23), Annex II: "Draft introductory statement for the World Plan of Action for the Application of Science and Technology to Development, prepared by the "Sussex group".

[3] ibid.

[4] idem: *World Plan of Action for the Application of Science and Technology to Development* (New York, 1971; Sales No.: E.71.II.A.18).

does not immediately reduce the imbalance of technological capacities in the world or reduce the technological dependence of the developing countries. Furthermore, it gives rise to the risk of an incomplete linkage between the creation of appropriate technology and its application in the actual production process in developing countries. At the very least, technology created in industrialised countries on behalf of developing countries requires a strong national counterpart within the latter for testing, adaptation, dissemination and training. It is in that sense complementary to the building up of national technological capacity in the developing countries rather than a substitute.

Perhaps the foremost reasons for placing great emphasis on the building up of national capacity to create appropriate technology in late developing countries are political. The state of technological dependence is dangerous and harmful to the late developing countries, permeating all international relations ranging from trade and consumption patterns to military power. It is not a natural or desirable condition for the global system and this is increasingly recognised in the industrialised countries also.

The chief way in which the creation of national capacity to develop new technologies will affect a country's technology mix may be defined as follows:

(1) Assuming that the country's development policy is firmly oriented towards the reduction of poverty, technology can be directed towards those goods and services which form part of the basic needs of the community. These "appropriate products" are likely to be different from the products developed in the rich countries. Where this is the case, no process of selection of existing technology, however careful, will serve the poorer countries.

(2) The factor endowment of developing countries requires a technology that is different to a degree which may be beyond the range of the available spectrum. The principle of appropriate technology in any country must be based on maximising the productivity of the scarcest factor of production. In industrialised countries this leads to a principle of maximising labour productivity. In developing countries the principle must be to maximise the productivity, or output per unit, of capital, scarce skills and of land (where land is scarce). The increase of labour productivity is, of course, the ultimate goal of development; but the best way of reaching this goal is through maximising the productivity of capital and available skills until these cease to be the scarcest factors of production.

Although the objective is to create a national or indigenous technological capacity, it does not follow that this is best achieved by separate efforts on a purely national basis. The expenditures involved are often beyond the means of individual low-income countries. An industrialised country may spend 3 per cent of its gross national product on R and D and related expenditures; a matching effort by a low-income country would be out of the question. Moreover, the results of such expenditure are risky since any given new technology may be affected by subsequent technological developments; they are uncertain since the objective prevents any simple input-output relation; and they are intangible since the value of the new technology depends on its adoption in the production process, which in turn depends on a number of factors beyond the expenditure of resources to create new technologies. This obviously is a fruitful field for co-operation among developing countries (and if possible also the industrialised countries), specifically including an exchange of information and results as well as national specialisation within an agreed wider framework.

In spite of the distinctions between these three different approaches to technology, in some ways the successful assimilation, adaptation and creation of new technology all have certain requirements in common (such as management skills and needs for skilled labour, training, and so on, as well as institutional and organisational requirements). These are discussed in Chapters 6 and 7 below.

The three "supply factors" (that is, transfers of existing technology, adaptation and new technology) will be considered in more detail in later parts of this study. Now the "demand factors" identified above may be briefly considered.

THE STATE OF FACTOR PRICES AND OTHER INCENTIVES FACING DECISION MAKERS

There is one school of thought which would maintain that there is little or nothing wrong with technology. If only, it is thought, the developing countries would remove factor price distortions, especially the overvaluation of unskilled labour in the modern sector and the undervaluation of capital and foreign exchange, appropriate production techniques would automatically be selected. This is an extreme formulation, and not all those supporting the emphasis on factor price distortions would go as far as to present it as a sufficient answer to the problem of inappropriate technology.

This view, which may be characterised as a market-oriented (liberal) approach, is based on several assumptions:

(1) There is a wide spectrum of technologies available from which, with appropriate factor price ratios, the more capital-saving technologies will be selected. The opposite assumption is that of "fixed" or "rigid" factor proportions with little "spectrum". Recent research has tended to confirm that a spectrum does in fact exist, although this does not necessarily mean that it is broad enough at its labour-intensive end to make the existing technology appropriate for the employment needs of developing countries, even with factor prices offering full incentives for the exploitation of capital-saving opportunities.

(2) Additional technological flexibility in response to factor prices and demand factors is provided by the possibilities of quality differentials

within sectors. Furthermore, technological flexibility does not only imply the use of different technologies within a given sector or to produce a given product, but may also be achieved by shifting production from relatively capital-intensive products or sectors to more labour-intensive products or sectors. Such shifts may also be induced by an appropriate price system.

(3) The price system of developing countries is taken into account by those in a position to decide on the development of technologies. In so far as the existing technology is created in and for the industrialised countries, this is obviously a very doubtful assumption. Furthermore, as regards investments made by multinational enterprises it must be assumed that the technology created at their head offices and central R and D establishments is little influenced by factor prices in developing countries. The assumption is, however, much more likely to be true for adaptations made within a developing country or for new technologies specifically created in and for developing countries.

(4) Factor prices are supported rather than negated by other incentives. It would be clearly ineffective to create pressures for appropriate technology through appropriate or shadow pricing if at the same time the pressures were negated by tariff protection, fiscal policy, profit guarantees, and the like.

(5) There is no doubt that factor price distortions (for instance, high wage rates in the modern sector) do co-exist with inappropriate technology, but this does not necessarily prove that distortions actually cause inappropriate technology. The line of causation may run otherwise. Thus the inappropriate technology in the modern sector, with its privileged access to capital and high labour productivity, may be the cause of the relatively low cost of capital and the relatively high level of wages in that sector. In this case, it is not the correction of factor prices which is the indicated policy, but direct action in the field of development policy and income distribution. This is the view of those who emphasise the importance of institutional factors rather than market forces in the determination of factor prices.

(6) The impact of factor price distortions on employment is not always clear-cut. The dualistic structure of the economy which results from inappropriate technology (that is, capital-intensive technology in countries with severe capital shortage) also serves as a form of protection for the small-scale and informal sector. A reduction of wage rates in the modern sector, with a consequential reduction of the protection of the informal sector, should therefore be combined with policies to upgrade and assist the small-scale and informal sectors, and not be considered as the main cure of the employment problem.

(7) Most important, the emphasis on removing distortions in factor prices is based on the assumption that those making actual decisions on tech-

nology are in fact influenced by price and profit calculations. The same researchers who showed that there is a labour-intensive spectrum in existing technologies have also been puzzled as to why this end of the spectrum is neglected in spite of the absence of any apparent obstacles to profitability. One plausible explanation[1] is that decisions on the technologies which are included in feasibility studies or preferred in the final choice may be influenced by engineers or others who, by their training, background and orientation, are inclined to consider the latest or most modern technology, or the one maximising labour productivity, as inherently and automatically superior. The case studies have clearly brought out the importance of such maximising considerations as against marginalist economic influences. Similarly, multinational enterprises deciding on technology for investments in developing countries may prefer capital-intensive technologies which reduce the need for what are, for them, unfamiliar supervisory skills. The choice of technology for public projects such as road construction may be influenced by the higher speed of capital-intensive methods. The more rapidly a job can be finished, the less need there is for supervisory skills, the housing and feeding of a large labour force, or to carry budgetary funds over into new budgetary periods. All such considerations reduce the influence of prices.

The studies undertaken under the ILO World Employment Programme do not give clear support to those who would put forward the relative prices of labour and capital as the main determinants of technological choice. For India it has been concluded that: "Reasons 'beyond factor prices' seem to be important even in market-oriented economies. The Indian experience with the river valley projects suggests that considerations of scale, time, location and productivity considerably reduced technical flexibility."[2] Case studies relating to can manufacture in Kenya, the Indian sugar industry, the Mexican metalworking industry and copper mining in Zaire, Zambia and Chile[3] do in general indicate that other more technical factors are at least as important as the relative prices of labour and capital.

In the case study of can manufacture it was found that the wage rates at which more labour-intensive technologies would become profitable would have to be unrealistically low. The Indian sugar study brought out the influence of government controls and other factors such as trade union pressure against minimisation of labour costs (both in this case operating in favour of the small-scale labour-intensive producers) which are more important than wages or other price factors. Moreover, it was shown that in a processing industry such as sugar, because of different recovery rates, the

---

[1] See Pickett et al., op. cit.

[2] A. S. Bhalla: "Technological choice in construction in two Asian countries: China and India", in *World Development,* Mar. 1974, p. 73.

[3] See ILO: *Technology and employment in industry: A case study approach,* edited by A. S. Bhalla (Geneva, 1975).

23

essential price factor is that of the raw material (sugar cane) rather than capital and labour. In extracting industries too, the characteristics of the raw material often determine the appropriate technology. In the Mexican metalworking industry, on the other hand, capital cost factors such as the cost of factory space in urban centres (but also the size of the production batch or lot) were shown to be the governing factors. Elasticities of substitution between labour and capital in copper mining in developing countries were shown to be low (in fact, they were lower than in the United States).

The general conclusion is that, while factor prices may be useful in guiding producers to a choice of certain technologies, their importance is often outweighed by other factors. Thus, the correction of any distortions in factor prices does not seem in practice to have quite the importance often attributed to it by economists. But it must be reiterated that the research results are by no means unanimous on this.[1] Sometimes, of course, the price factors may be present but inextricably mixed with other effects: for instance, production on a smaller scale in developing countries often leads to more labour-intensive methods and at the same time the wage rates in the smaller-scale sector are usually lower. But it is not easy to disentangle these two effects. Those who have made practical studies of the situation seem to assume that it is the technical requirements of the scale factor rather than the lower wages which induce the more labour-intensive technology.

To sum up: it is not open to question that the establishment of a system of prices designed to reflect factor scarcities, or the use of "shadow prices" in project preparation and evaluation,[2] can be a useful and perhaps indispensable part of any effective technology policy. Equally clearly, however, there are a number of inherent limitations in this approach which prevent it from being acceptable as sufficient in itself—that is, as a substitute for more direct technology policies. In fact, the practical case studies would suggest that it is far from being the major part of total technology policy.

EFFECT OF INCOME DISTRIBUTION

Both the strategies mapped out by the ILO comprehensive employment strategy missions and the results of empirical case studies have served to identify income distribution as a key determinant of technology. For a basic-

---

[1] For a different result highlighting the importance of factor prices and warning against underrating elasticities of substitution, see Howard Pack: "The substitution of labour for capital in Kenyan manufacturing", in *The Economic Journal* (London, Cambridge University Press), Mar. 1976.

[2] See W. A. McCleary et al.: *Equipment versus employment: A social cost-benefit analysis of alternative techniques of feeder road construction in Thailand* (Geneva, ILO, 1976), especially Ch. 3; G. W. Irvin et al.: *Roads and redistribution: Social costs and benefits of labour-intensive road construction in Iran* (Geneva, ILO, 1975); and Deepak Lal et al.: *Men or machines: A study of labour-capital substitution in road construction in the Philippines* (Geneva, ILO, 1977).

needs strategy it is clear that products have to be selected for their suitability for low-income consumers and low-income producers. Hence the emphasis on "appropriate products" as a precondition for "appropriate technology". The empirical case studies have clearly brought out the importance of the nature of the product and of product specification and product quality for both optimum and actually preferred technology. The close connection between income distribution and nature of products has already been indicated earlier. Where the income distribution is unequal and where those at the upper income levels are given privileged access to scarce resources, the products in demand will be similar to those produced in industrialised countries and will therefore tend to have an established, basically capital-intensive technology. Once the unequal income distribution is firmly established, it is difficult to shift to more appropriate products since these would lack a sufficiently large market. In this sense, policies regarding income distribution should logically precede policies affecting products and technology.

A change in the nature of products towards more appropriate products, especially by increasing employment, will help to create and support a more equal income distribution. This offers possibilities of mutually reinforcing policies in the field of income distribution, on the one hand, and products and technology, on the other.

The circular relationship between technology and income distribution can also be expressed in terms of a change in the relative scope of the major decision-making groups in the economy. With a shift to a more equal income distribution which favours appropriate products, the scope of decision makers in multinational enterprises or very large national producers with a natural tendency to select modern technologies as more suitable to their purposes will diminish, while the scope of decision makers of smaller units with a natural tendency towards a more labour-intensive, more adapted technology will increase. Governments will also be induced to shift from providing the infrastructure required by large-scale technologies towards providing the infrastructure required by the intensified production of goods representing the basic-needs standard. These tendencies will be further strengthened by the fact that a more equal income distribution is likely to benefit the rural sector, so that the pattern of demand will tend to become more small-scale and decentralised. This, in turn, further favours the more labour-intensive indigenous small-scale technologies.

However, both the empirical case studies and economic analysis have brought to light some complicating features. The report of the ILO comprehensive employment strategy mission to Kenya embodied a dynamic model (admittedly based on assumptions rather than data) where an initial income redistribution raises the incomes of the poor, who as a result consume more simple goods produced by labour-intensive methods, thereby in turn helping more of the poor to find productive employment. A model developed under the World Employment Programme has been applied to the Philippines with similar results and is being used to analyse the effects of income redistribution

25

in Iran and Malaysia.[1] It is not invariably true that products consumed by the rich are more capital-intensive than those consumed by the poor. Exceptions range from the demand of the rich for domestic service, hand-made clothing, rugs and leather shoes, and so on, to the advantages of plastic shoes (more capital-intensive than leather shoes, but also cheaper and more durable), chemical fertilisers, bicycles, and so on, for poor consumers. As a generalisation, the picture of a complementary relationship between more equal income distribution and a shift to more labour-intensive products and techniques probably remains correct (especially where shifts between the modern and the informal sectors are involved); but the results of the case studies present a warning that it is necessary to study specific cases rather than to rely on general economic tendencies.[2] They also open up new possibilities for constructive technology policy.

Once a desired target towards more equal income distribution has been attained and the production mix reflects the necessary transition to more appropriate products, the role of technology policy is by no means finished. Technological improvements will be needed to raise over-all labour productivity and make it possible to raise the basic-needs standard. There will still be the need to adapt transferred technology as well as to upgrade existing national and small-scale technology. Moreover, the change towards more appropriate products and more local control and management which goes with a more equal income distribution can normally be expected to stimulate local innovation and to increase local R and D activities.

NATURE AND SITUATION OF DECISION MAKERS

The decision makers in matters of technology in developing countries represent highly diverse interests and motivations. They include multinational enterprises whose main objective is to maximise global profits within the whole multinational complex; large-scale modern sector national producers who basically wish to maximise profits but who are also influenced by other considerations; national governments who have specific development objectives, possibly including up-to-date modern technology and other aspects of modernisation, as well as the reduction of poverty; family enter-

---

[1] For detailed country-specific analyses, see especially ILO: *Employment, incomes and equality . . .*, op. cit., and idem: *Sharing in development: A programme of employment, equity and growth for the Philippines* (Geneva, 1974), and the subsequent Philippine modelling exercise (F. Paukert et al.: *Redistribution of income patterns of consumption and employment: A case study for the Philippines* (Geneva, ILO, 1974; mimeographed World Employment Programme research working paper; restricted). See also ILO: *World Employment Programme: Research in retrospect and prospect* (Geneva, 1976), pp. 32 and 128.

[2] Recent simulation exercises "show that even radical redistribution experiments affect the sectorial composition of gross output only modestly, and that resulting indirect effects on importation and on capital and labour use are correspondingly modest". C. F. Díaz-Alejandro: "North-South relations: The economic component", in *International organization* (Madison, Wis., University of Wisconsin Press), Winter 1975, p. 215.

prises such as smaller farms or small production units in the informal sector whose objective is to maximise family consumption or output per unit of land. Within each of these groups there is a further diversity of interests. For instance, engineers in modern sector firms may base their technological choices on principles different from those of commercial managers; thus the actual choice of technology may depend on the relative sphere of influence of engineers as against commercial managers (although the evidence concerning the respective positions taken by these groups is contested). Within the public sector too, different decisions on technology may be made by different government departments controlling different sectors of the economy.[1]

Reality is more complicated than any simple and purely economic model in which the choice of technology results, in a clear and straightforward manner, from the confrontation of factor prices and the principle of profit maximisation. The real situation is deeply influenced by the distribution of economic, social and political power between the different types of decision maker. This creates the need for some kind of political economy of technological choice—a field very much in need of further exploration on the basis of additional empirical data. But the present study has already served to emphasise the importance of this factor, some aspects of which will be discussed in subsequent chapters.

---

[1] See C. Cooper et al.: "Choice of techniques for can making in Kenya, Tanzania and Thailand", in ILO: *Technology and employment in industry,* op. cit.

EXISTING TECHNOLOGY: SELECTION, TRANSFER AND DISSEMINATION

3

SELECTION

Even though the present distribution of the world's generation of technology—as broadly measured by R and D expenditures—may gradually become less one-sided, and even though the developing countries will gradually do more to adapt and upgrade their own technologies, for the foreseeable future the transfer of technology from developed countries will remain a major source of technology for the developing countries. Its appropriateness in the light of a basic-needs strategy and other legitimate development objectives (including the quickest possible reduction of technological dependence) will therefore be a development problem of the highest importance. The appropriateness of this major source of technology and its value for development will depend on proper selection, proper conditions of transfer and proper dissemination within developing countries. It is also important to combine the imported technology with specific domestic technology that has been developed either nationally or by those controlling the original technology, and to adapt it to the special conditions prevailing within each country.

As has been pointed out, empirical research has confirmed that for any given product there is usually a range of technologies available, and a further range may be opened up when appropriate product variations are included in the development strategy, in the light of the basic needs of the population. However, in practice the theoretical or potential range of technologies available is narrowed down by a number of factors.

Among these may be mentioned the lack of available information. Obviously, only the *known* range of technologies can be effective, and knowledge or information in respect of existing ranges of choice requires skills and capacity and/or proper institutional support as well as financial resources. This lack of information is particularly marked among small-scale and informal sector producers and among small farmers. Even where relevant extension services exist, the small producer usually finds that he has very much less access to such services than the larger producer. Moreover, such services are often specifically geared to the products and problems of the larger producer.[1] This has been a theme of a number of the ILO employment strategy missions. The work carried out as part of the World Employment Programme and elsewhere has shown that the lack of information or awareness of alternative technologies is a key problem.[2] In particular, it makes it more difficult for developing countries to obtain technology *per se*. To do so would often be more appropriate than to obtain it in packaged form as part of investment and management arrangements or embodied in imported machinery and equipment.

Again, there may be no incentive to use appropriate technology. The choice of technology is made by a multiplicity of decision makers, each of whom has his own objectives and is subject to different pressures. Multinational enterprises will tend to apply their own technology, as developed by their own R and D in their home country; indeed, to gain increased benefit from the R and D expenditures already incurred may be one of the incentives for investing in a developing country. A machinery salesman will wish to sell his own brand of machinery and has no incentive to discuss other brands. Engineers and engineering consultants, by their training and background, will often identify appropriate technology with the maximisation of labour productivity (which will normally lead to highly capital-intensive technology and to technologies with high skill requirements). Enterprises which, in pursuit of a policy of import-substituting industrialisation, enjoy virtually guaranteed profits through tariff protection or other subsidies will have no special incentives to expend much effort on finding the most appropriate technologies. Governments, through permitting distorted factor prices and through fiscal and other incentives, may give incentives for the choice of inappropriate technology; even in their own public investment there may be a failure to apply the kind of social cost/benefit analysis giving full weight to employment creation and reduction of poverty which would lead to appropriate technology. The ILO comprehensive employment strategy missions and other studies have brought to light much evidence of such lack of incentive. Governments may also bring about the selection of inappropriate technology through a failure—often due to limited budgetary resources—to provide information services for decision makers. In the case of projects financed with the help of tied aid, the technology may be limited to that preferred by the donor country and embodied in the equipment supplied by it. The educational pattern of engineers, scientists and administrators may also give rise to "blind eyes" for part of the technological range.

This catalogue of examples of the lack of information or the lack of incentive to use appropriate technology could be readily extended, but it will be sufficient to indicate the wide gap in real life between the theoretical existence of a wide range of possible alternatives and the frequent practical

---

[1] The same applies also to methods of dissemination (see the section on "Dissemination" below).

[2] This lack of information may affect even multinational enterprises: see Cooper et al., op. cit., p. 112.

absence of anything approaching a free and rational choice from this theoretical range.

So far, the reasons for the widespread failure to select appropriate technologies have been listed. It must however be emphasised that the criteria of what constitutes "appropriate" technology can be discussed only in the light of a country's development objectives. The present discussion is conducted on the assumption that the objective is to increase productive employment within the framework of a basic-needs strategy. In the past, other objectives such as an increase in gross national product or the rapid "modernisation" of the economy have been predominant. Indeed, changes in development objectives are constantly occurring. Although employment is now recognised to be a major element in a basic-needs strategy, new concerns such as environmental factors and the need for oil-saving technologies will bring about modifications in technological choice. The effect of such changing objectives and conditions, as well as the difficulties in the way of selecting appropriate technologies, is that the actual technology in use will not be optimal. There is thus a need for a constant critical examination and diagnosis of the technology used both in continuing and in new projects. Such studies could be extremely useful for the determination of future policy, for the lessons of the past can serve to suggest ways in which existing technologies may be adapted. The ILO comprehensive employment strategy missions did not concentrate exclusively on technology, and in any case the length of time required for intensive studies of the kind envisaged would be much greater than the duration of any external mission. These studies should be conducted by national governments, but it would be natural for the United Nations to offer assistance in their preparation, execution, interpretation, dissemination and follow-up. Such studies would appear to be the most natural link between past experience and the formulation of better policies in the future.

The assessment of the technology used in a country calls for much more than a simple assessment of capital intensity. Many other factors have to be taken into account. An apparently highly capital-intensive or mechanised technology may in fact be less capital-intensive than an apparently labour-intensive one, if the machinery is used intensively. Moreover, the amount of fixed capital or machinery, on which those discussing the nature and choice of technology often concentrate, is only one aspect of capital: in most situations, working capital for raw materials or other inputs is very important, and it is frequently found that the apparently less capital-intensive or more labour-intensive technologies do, in fact, often have a higher ratio of working capital to fixed capital, as well as possible lower rates of utilisation. Thus a detailed assessment may reverse the classification of technologies by degree of capital intensity. [1]

A comparison of two technologies by capital intensity must also take into

---

[1] See Amartya Sen: *Employment, technology and development* (Oxford, Clarendon Press, 1975), pp. 47-48.

account the roles of other factors of production, such as land and other resources, training and education, entrepreneurship and scarce management resources, as well as the resources currently used in the production process under different technologies. Furthermore, when measuring the volume of employment created by different technologies, one must also take account of the importance of indirect employment, which economic analysis has shown to be often a high multiple of direct employment. Such indirect employment may arise from backward linkages with sources of supply, forward linkages towards processing and distribution, multiplier links through the expenditure of incomes created in the production process, and possibly foreign exchange multipliers arising from the differential contribution of different technologies to the balance of payments. It is perfectly possible that, when two technologies are compared, the one which is more labour-intensive in the sense of providing more direct employment will in fact turn out to be less labour-intensive when the indirect effects are taken into account. The indirect and wider effects of different technologies may be quite different. They need to be taken into account in the policy studies proposed above.

The actual criteria which should govern the selection of technology must differ for countries with different endowments, different objectives and different levels of development. Hence, desirable criteria can be listed only in a very general way and with the reservation that governments may well attribute different weights to the various elements in this list. For the purposes of formulating an effective technology policy it might be found useful for each country to make its own list explicit, and even attach some broad quantitative weight to the various criteria. Based on both economic analysis and the evidence of the ILO comprehensive employment strategy missions, case studies and monographs, such a list of criteria would normally include: *(a)* high employment potential, including indirect employment through backward linkages with national suppliers and forward linkages with national processors, distributors and users; *(b)* high productivity per unit of capital and other scarce resources; *(c)* higher labour productivity in the context of increased employment, that is, the maximisation of the productivity of labour in the economy as a whole, rather than for each individual project; *(d)* the utilisation of domestic materials, especially of raw materials previously considered valueless; *(e)* a scale of production that is suitable for the local markets to be served (unless exports are involved), with special consideration being given to small and fragmented markets in rural areas; *(f)* low running costs and easy and cheap maintenance; *(g)* maximum opportunity for the development, as well as use, of national skills and national management experience; and *(h)* dynamic opportunities for the further improvement of technologies and feedback effect on the national capacity to develop new technologies. For countries accepting and practising a basic-needs strategy, three obvious additions to this tentative list will be: *(i)* products which meet the basic-needs standard and are suitable for the target population that it is intended will benefit from a basic-needs strategy; *(j)* the generation of incomes for the poorer sections of the population, or of incomes which can be channelled towards the poorer sections; and *(k)* technologies which are suitable for production in poorer areas or regions. The application of such criteria is greatly facilitated when an industrial process is subdivided into its various operations and when the criteria are applied separately to each stage—another aspect of "unpackaging".

These criteria relate to the objective of maximising productive employment within the context of a basic-needs strategy. Governments will wish to add other criteria derived from other specific policy objectives.

Governments can affect the selection of technology through a wide variety of instruments (for example, their own selection of technology for projects in the public sector; their influence on the decisions of private investors through fiscal, credit, price, wage and other policies and incentives so as to encourage or discourage the adoption of certain technologies; the nature of public infrastructure (which can have considerable influence on the technology adopted in the private sector); their policies regarding import substitution and export promotion; and the operation of foreign exchange controls, licensing and permit systems). Governments can also provide direct advisory services which would draw the attention of private managers and producers to technologies that are considered more appropriate to development needs than those introduced by machinery salesmen, foreign investors or other current sources. Such services may be particularly valuable to smaller and informal sector producers.

One factor which has tended to be neglected in the past is what may be called the "politico-structural" factor. As has been seen, the criteria used for selecting technology, and hence the resulting technologies, vary widely according to whether the decision maker is a government, a multinational enterprise, a large-scale national enterprise, a small-scale national enterprise, a producer in the informal sector, a large farmer, a small farmer, and so on. The motivation of each of these decision makers is quite different from that of the others. It follows that a government can exert a very considerable influence on the shape of technology through the relative importance and scope given in the economy to the activities of these various categories of decision maker. Policies regarding the scope of the public sector, the encouragement of larger or smaller units of production, land distribution, the allocation of essential inputs between larger and smaller farmers—all these strategic decisions, although they may be governed by much wider forces than technology considerations, have a profound influence on the selection of technology.

The spread of "modern" consumption patterns, coming mainly from the industrialised countries with the help of advertising media and various cultural influences, including the "demonstration effect" of exposure (usually initially through imports by richer sections of the population or expatriate residents), is equally important for technological choice. Governments may reasonably adopt varying policies regarding those changes in taste which have both harmful and desirable aspects. In many cases it may be possible to satisfy the demand for "modern" goods without having recourse to "modern" technology, which will often be inappropriate for the developing countries. In other cases a government may wish to reduce or modify the demand for "modern" goods and try to control advertising and other means of creating new taste patterns. In the non-oil-producing developing countries the need to save foreign exchange and to turn to energy-saving forms of production is a strong incentive for the modification of consumption patterns (as well as for the adaptation and modification of technology). There is a complementary relationship in that technologies which are import-saving and energy-saving will often (although not invariably) also be those which are more appropriate for and preferable under a basic-needs strategy. Here, however, concrete case studies are needed, similar to those already undertaken regarding other aspects of technology.

The same is true of the relationship between the selection of appropriate technology and the increased emphasis on the export of processed and manufactured goods. In the attempts now being made under UNCTAD auspices to give the developing countries' exports of these goods easier access to the markets of the industrialised countries, some thought must also be given to considerations of technology and hence of employment. Here there are various factors working in opposite directions. The fact that the comparative advantage of the developing countries in international trade lies in labour-intensive industries and operations will naturally tend to emphasise labour intensity. A factor working in the same direction would be the greater competitive pressure in world markets which would induce producers to search actively for cost-saving technologies (and also to adapt existing technologies more actively). In the opposite direction, there is the fact that larger-scale firms have an advantage in entering export markets because it is easier for them to establish the necessary marketing contacts, do market research and guarantee the steady supply and quality of their products. Multinational enterprises with their international connections will hold a natural advantage which might strengthen the possibility of their introducing technologies that might reduce the employment effects in industries processing and manufacturing goods for export. Similarly, the necessity to produce high-quality products of a uniform standard appropriate to rich country markets will give an advantage to large-scale producers, and may put a premium on capital-intensive technologies.

In the absence of more concrete research and case studies in this field,[1] it is not easy to say how these various factors will balance. It is, however,

---

[1] A beginning has been made in journal articles, notably S. Watanabe: "Exports and employment: The case of the Republic of Korea", in *International Labour Review* (Geneva, ILO), Dec. 1972; C. Hsieh: "Measuring the effects of trade expansion on employment: A review of some research", ibid., Jan. 1973; and S. Watanabe: "Constraints on labour-intensive export industries in Mexico", ibid., Jan. 1974. See also H. F. Lydall: *Trade and employment: A study of the effects of trade expansion on employment in developing and developed countries* (Geneva, ILO, 1975).

perhaps more important to inquire what governments can do to strengthen the factors favourable to the development of appropriate technology in connection with the expansion of exports. For example, anything that can be done to enable small-scale producers to participate in exports, perhaps on a co-operative basis, would be helpful in this direction. In any case, the additional foreign exchange receipts resulting from a successful expansion of exports would make it easier for developing countries to acquire technology *per se,* rather than having to accept it as part of a package with investment, supplies of machinery on a loan basis, or management services. If export earnings freely available to developing countries replace tied aid, this should *pro tanto* increase the freedom of developing countries to obtain appropriate technology.

TIED AID

Tied aid has a clear disincentive effect on the selection of appropriate technology. It puts a premium for the developing countries on minimising local employment, the use of local materials and other local cost items (which do not attract aid) and maximising the use of imported equipment (which does attract aid). Thus tied aid not only reduces the freedom of the recipient to choose the most suitable equipment available in various countries, but even more fundamentally it may distort the aided project in the direction of greater capital intensity. In addition, with the resources in developing countries to formulate and develop projects being limited, tied aid will divert these scarce resources towards the formulation of more capital-intensive projects which attract aid, and away from labour-intensive projects which do not. There is no doubt that tied aid is inherently antagonistic to appropriate technology. Donors have very little to lose if they untie their aid by multilateral agreement or (even better) put it on a programme or budgetary basis. The developing countries will still use their additional foreign exchange receipts for additional imports, largely from the industrialised donor countries. Failing such a multilateral donor agreement, it will be more difficult for individual donors to untie unilaterally, especially when they are faced with domestic unemployment and balance-of-payments problems.

Similarly, the fact that aid is normally not available for the recurrent costs of existing projects but only for "new projects" is a drawback in the establishment of appropriate technologies. The better utilisation of existing equipment will almost always be more labour-intensive than a new project based on new equipment. There is some evidence of widespread under-utilisation of equipment in developing countries,[1] possibly to an even higher degree than in industrialised countries. Where existing equipment is idle for lack of

---

[1] G. C. Winston: "Capital utilisation in economic development", in *The Economic Journal,* Mar. 1971.

foreign exchange to import raw materials, spare parts or maintenance equipment, it is highly likely that aid with recurrent costs would be more useful than aid for new projects. Those concerned with selection of appropriate technology in developing countries would certainly wish the OECD Development Assistance Committee to move towards multilateral liberalisation of aid practices in such matters as tied aid, the financing of local and recurrent expenses and greater emphasis upon programme and budgetary aid.

If tied aid is given it would be far better if it were tied to the development of national scientific and technological capacity inside developing countries through supplies of equipment for laboratories, technical assistance experts, the provision of libraries and technical reference services, essential R and D work in the donor countries' own laboratories, arranging visits of scientists and technologists, training facilities, services to developing country institutions through "twinning" arrangements (i.e. establishing a close link with intensive exchange of know-how and personnel between two specific institutions in an industrialised and a developing country respectively), and so forth. This form of tied aid—tied to the building up of the capacity to use local labour and local materials for the production of local products—certainly seems preferable to what is usually understood as tied aid.

Certain advantages are claimed for tied aid. The tying to projects is supposed to be a good basis for technical assistance, for demonstrating the importance of good project preparation, for building project preparation capacities among planners and technicians, for giving an incentive for keeping proper time schedules and quality standards in the construction of capital projects, and so on. However, all these advantages could perhaps be achieved by technical assistance and arrangements other than tied aid. In any case, all these qualities are needed as much for non-aided as for aided projects, and there is a risk that the concentration of efforts on the aided projects is at the expense of the non-aided.

A concrete case study of feeder road construction in Thailand[1] showed that the bias in the direction of increased capital intensity imparted by tied aid (especially tied grants) was considerable. Conversely, the untying of aid would have resulted in considerably greater labour intensity. This was so although other factors, such as methods of contracts and bidding, were also found to be significant determinants of technology, and in turn related to aid practices. The study concludes: "one must ask whether the merits imputed to project-tied aid are worth the sacrifice in employment that would appear to stem from a switch to untied aid".[2] One may add that aid practitioners as well as economists have been aware for some time of the illusory element in project tying, due to the existence of "fungibility" in the over-all allocation of investment resources.

---

[1] McCleary et al.: *Equipment versus employment...*, op. cit.
[2] ibid., p. 71.

TRANSFER PROBLEMS

Both the definition and the statistical measurements of the transfer of technology (and hence also the measurement of the cost and benefit of transfers) are made difficult by the fact that the transfer of technology takes many different forms. The transfer of technology *per se* (of designs, use of patents, blueprints, technical literature, technical advice) is but a relatively small part compared with other forms of transfer in which technology is combined (in a package) with other things in which it is embodied. The purchase of any machine, fertiliser, truck, indeed any commodity, not only embodies the technology by which it has been produced, but also requires a technology of application, use, maintenance, and so on. Although usually one does not stretch the term "transfer of technology" to cover the sale of, say, a motor car, there is good reason for the emphasis given by some analysts to the fact that the import even of consumption goods which embody the result of a certain technology requires a whole set of other complementary facilities, which in turn will determine the local technology. This is clear in the case of motor cars which require a network of roads of a certain standard, petrol filling stations, garages for repair and maintenance, and the like. In other cases, the impact of the embodied transfer may be less visible and more subtle but present all the same.

It is, however, true that the technology embodied in the motor car does not affect the developing country importing it in the same way as if the motor car had been domestically produced. Whether the motor car has been produced in a relatively more capital-intensive or labour-intensive way does not affect the impact which the motor car has on the economy of the importing country. The income and other economic and social effects of the technology used in the production of the car are felt in the country where the motor car is made, not in the importing country. This also applies to the learning and training effects. There is therefore some good reason to exclude the transfer of motor cars from a definition of transfer of technology. On the other hand, with machines the impact on the technology of the importing country is so direct and obvious that the technology embodied in the machine is by general agreement included in the concept and analysis of transfer of technology.

Even so, it is worth emphasising that the difference between a machine and a motor car is a difference more of degree than of kind. In particular, it is questionable to exclude the motor car on the grounds that it is "consumption" whereas the machines are "production". There is a technology of consumption as well as a technology of production. Control of the technology of consumption can be an indispensable part of technology policy. This is indeed recognised in the Report of the Director-General to the ILO World Employment Conference.[1] A basic-needs strategy is a form of control of the technology of consumption, among other things.

---

[1] ILO: *Employment, growth and basic needs...,* op. cit.

Economic aspects

Among the most important packages which contain technology as an element of transfer are those which take the form of a complete "turn-key" plant which embodies the work that has gone into the design and selection of the component parts and processes, the construction, purchase and financing, and so on. As another layer on the wrappings of the technology package, the finished plant may be financed and operated by an enterprise from an industrialised country (often a multinational enterprise[1]) as a branch or subsidiary, or it may be operated by a local agency under licensing agreements. In these cases the local powers of adaptation are also limited— unless, of course, such local powers are exercised by the government of the developing country as part of the conditions for admitting the foreign investment or approving the licensing agreement. Payments for transferred technology are contained in the accounts of commercially oriented enterprises with a tax and ownership base in other countries, and this makes the assessment of the cost of transfer of technology an immensely difficult task.

For this and many other reasons the developing countries feel that it would be to their advantage to move from packaged imports of technology towards imports of technology *per se,* that is, in unpackaged form. This form of transfer gives the developing country greater scope to combine technologies from various sources, to exercise freedom of choice, to adapt technology to local uses and to benefit from the learning and training process of applying and adapting technology. However, as has been pointed out before, there is a complementary relationship between the effective selection of technology from abroad and national technological capacity. The greater flexibility and increased power of adaptation which the unpackaging of technology transfer could provide can actually be used only by countries which have the technological capacity and skills to do so effectively. Un- packaging alone would be least useful to the least developed countries.

As long as the possession of technological power is unevenly distributed in the world, even the transfer of technology in unpackaged form could still be very costly to developing countries, as well as being accompanied by restrictions. Among these restrictions is a prohibition on sharing the technical knowledge with other developing countries and with other companies in the same country. This inhibits the spreading of technical knowledge and also creates a costly need to import the same technology over and over again. Consequently, unpackaging would not by any means solve all problems, and the objective of making technology available to developing countries on reasonable terms and reasonably free from restrictions on use could be achieved only by international agreement, for instance by an effective code for the transfer of technology (such as is now under discussion) and the reform of the patent system. Even if it is assumed that agreement can be

---

[1] For a discussion of wider aspects of foreign investment by multinational enterprises, which is not attempted here, see ILO: *Employment, growth and basic needs...,* op. cit.

reached, payments by developing countries in various forms, both open and concealed, for technology transfer will continue to be a heavy, and probably increasing, burden.

Some of the fees received from developing countries through patent and licensing arrangements,[1] as well as some of the profits of multinational enterprises derived from developing countries, serve to strengthen and expand R and D activities in the industrialised countries, and so help to perpetuate the global technological imbalance. In order to achieve a more even spread of national technological capacity, any resources released by changes in the terms of transfer negotiated by developing countries could be channelled into indigenous R and D.

One problem is that the international transfer of technology is almost entirely dominated by large companies with considerable R and D establishments. However, the technology which would often be particularly useful to the developing countries is that developed by medium- and small-scale enterprises in industrialised countries. This potential remains largely untapped. Such enterprises might be ready to share their technology with developing countries and to impose fewer restrictions, since they are less concerned with world-wide operations and international feedback effects. Here is another promising area for future international action.

There is also an internal transfer problem within developing countries, in that there exists a wide range of technologies of varying productivity, even among smaller enterprises which are not inhibited from sharing their technology by any restrictive conditions under the terms on which it was acquired. The task of making such technologies available could be taken on by the national research and technological institutions of developing countries, especially those which are technologically more advanced and have a large number and variety of industrial enterprises.

Social aspects

There is another aspect of the "negative transfer" of advanced technologies, namely the undesirable consequences of transfers from industrialised countries on working conditions and the working environment in developing countries. The increasing awareness of the disruptive effects of technology has led to a growing dissatisfaction with narrow assessments that do not take sufficient account of the social and human consequences of technology transfer, and to a growing recognition that these factors may prove as important as economic and employment factors. There are however some indications of a new approach to this problem, with economic, employment and other social goals being examined together.

Quite often the social costs of transfer are either subsumed or made subservient to economic costs. For example, the social repercussions of

---

[1] As was noted in Chapter 2, the cost of these alone was estimated to amount to at least US$1,500 million by the end of the 1960s, and to be increasing by 20 per cent per annum.

modern technology may take the form of longer hours of work, increased fatigue, job alienation, mental stress, environmental pollution and an increased risk of occupational hazards and accidents. Not enough is yet known about the constraints that recent technological advances, including automation, place on job design, work organisation and work experience.[1] Technology transfer does not involve merely the transfer of equipment: it also brings with it such organisational techniques as the routine assembly-line methods associated with the moving conveyor belt in use in most manufacturing industries in the industrialised countries. Such methods may not be very conducive to worker flexibility in controlling work pace. Experiments have been carried out in breaking up long conveyor belts into shorter sub-assemblies and product-oriented work groups, each responsible for a series of tasks. This reorganisation of work and production tends to reduce the monotony of mass production. Such methods can be adapted to the particular cultural patterns and skills of the developing countries.

New methods and new patterns of organisation of work can create psychological problems of adjustment for workers who may not have been exposed to the industrial environment that is necessary for the proper implementation of these work patterns. The pace, discipline and hierarchy of industrial life may be totally different in different cultural environments. What kinds of cultural value facilitate changes in attitudes? This would be an interesting area for investigation. The experience of Japan suggests that technological changes can be brought within the framework of a national culture. Can the same be achieved in other countries?

Under the ILO International Programme for the Improvement of Working Conditions and Environment (PIACT) studies are already under way in Algeria and Mexico on the relationships between conditions of work and technology transfer. It is intended to extend these studies to other countries. Information is being collected on the extent to which imported technology and forms of work organisation have affected conditions of work and life in developing countries, the reactions of individuals and social groups to the introduction of imported technology, and the efforts that have been made or should be made to adapt technologies to local climatic, cultural and social conditions in order to minimise their harmful side-effects. It is also proposed to undertake a comparative country analysis and develop a methodology for the evaluation of alternative technologies, taking into account social effects such as those on working conditions and environment.

The transfer and use of second-hand instead of sophisticated machinery can also have negative social consequences. If such machinery is inadequately guarded or maintained, health and safety hazards will be created, even though its use may be otherwise desirable for economising capital and

---

[1] See ILO: *Technology for freedom: Man in his environment—The ILO contribution,* Report of the Director-General/Part 1, International Labour Conference, 57th Session, Geneva, 1972. See also Jack Baranson: "Diesel engine manufacturing: De-automation in India and Japan", in ILO: *Automation in developing countries* (Geneva, 1972).

generating employment. However, machines without proper guards can be satisfactorily equipped at no great expense. Guidelines for ergonomic speci-fications in connection with the importation of modern technical equipment and installations can also be drawn up. In the early 1960s very few countries had any legislation requiring machinery manufacturers and vendors to put only properly guarded machinery on the market. It was for this reason that in 1963 the ILO adopted international labour Convention No. 119 concerning the guarding of machinery. Although this Convention has now been ratified by over 30 countries, very few of these are exporters of machinery; its effectiveness is consequently limited. [1]

Again, advanced technology such as continuous flow equipment yields a higher return when it is operated round the clock; yet its use entails night work, which may disrupt family and social life. [2] In some societies and cultures, workers are quite opposed to the idea of night shifts and overtime. Even when such constraints do not occur, it may be necessary to minimise the disruptive influences of night work and of fatigue due to overtime.

Although the two issues of the cost of technology transfer and the selection of appropriate technology are dealt with separately here, they are in fact inter-related. To lower the cost of the transfer of inappropriate technology would be of doubtful benefit to developing countries if as a result such transfers increase. The question of selection becomes even more important with any progress that may be made in increasing access to existing technology. The real answer to both problems is to strengthen the screening and negotiating capacity of developing countries.

DISSEMINATION

In addition to the problem of whether or not appropriate technologies are available for transfer, and whether the conditions of transfer are reason-able or not, there is the further problem of whether—and how—the knowl-edge of available technologies can be communicated to developing countries. These two distinct problems have been identified respectively as the "suita-bility gap" and the "communications gap". [3] The simple statement that communications must be improved if the selection of appropriate technology is to become more effective conceals some very difficult problems. The potential recipients of this information cover a wide range, including govern-ment departments and agencies, large and small enterprises in various sectors of the economy, R and D institutions, financing agencies, informal

---

[1] ILO: *Making work more human: Working conditions and environment*, Report of the Director-General, International Labour Conference, 60th Session, Geneva, 1975, p. 26.

[2] See J. Carpentier and P. Cazamian: *Night work* (Geneva, ILO, 1977), pp. 43 et seq.

[3] Paul Streeten: "Technology gaps between rich and poor countries", in *Scottish Journal of Political Economy* (London, Longman), Nov. 1972.

sector producers, consultants, and so on. The nature, scope and level of information needed by these various recipients differs greatly, as does their capacity to absorb and use information. Equally varied are the sources of information, partly abroad and partly domestic, again including government agencies, enterprises, financial and development institutions, R and D institutions, consultants, publishers of technical literature, sellers of equipment, and so on. To establish a comprehensive mechanism that can cover such heterogeneous sources and users, and provide information which can be made operationally useful at such different levels, is obviously a very difficult task.

It seems clear that the setting up of a comprehensive system must be approached in stages. The various sources and users must be individually studied and approached as to their willingness and ability to supply and use information. There is even a second-degree problem in communicating with potential sources and users of information about the existence of an information system and of its potential value. The use of information may also require special training; perhaps even suppliers of information would also need training. But in spite of the formidable nature of the task, the potential value of such a system requires much greater activity in this field, including international action. It is particularly important to remedy the neglect of the traditional, informal and small-scale sectors in existing information systems. This has been recognised in United Nations General Assembly resolution 3507 (XXX) concerning institutional arrangements in the field of the transfer of technology, and the subsequent report of the Inter-Agency Task Force on Information Exchange and the Transfer of Technology, discussed in the final section of this chapter.

The type of information needed extends beyond technical specifications and blueprints to essential economic data regarding cost of investment, maintenance cost, running cost, availability of needed raw materials and spare parts, cost effects of variation in scale of production and quality of production, skill and training requirements and requirements for supporting services. There is clearly a limit as to how much of all this should be covered by a technology communications system, and how much should be left to the potential user to find out for himself, if necessary with specific assistance. However comprehensive the information, it can never fully cover all the requirements of each specific case, and additional assistance will often be necessary.

To build up reliable, comprehensive and up-to-date services is obviously a major undertaking and will be quite costly. This is obviously an area where international co-ordination and assistance would be particularly valuable, and there is both need and scope for regional and inter-regional collaboration among developing countries. It would seem that, in spite of the high cost, the benefit/cost ratio of such a scheme would be quite high, considering the costly deviations from appropriate technology which now occur both for lack of information and as a result of sales pressures from commercially oriented sources.

The cost could be at least partly recouped by charges to the beneficiaries, but on this question there is room for dispute. In view of the tremendous economic and social cost of inappropriate technology to the community as a whole, it could well be argued that such a service should be free so as not to discourage any potential users. On the other hand, one wants to discourage the frivolous use of a service the resources of which are likely to be strained. Perhaps the solution is to make a charge where information provided by the service has been used with demonstrable private benefits to the user. However, this introduces a somewhat arbitrary element into the assessment of fees. The best solution might be to cover the cost by a general charge on industry as a whole.

At present, some of the needed services are rendered on a private basis by consultants specialising in the technical problems of a specific industry. Where the consultant is unbiased and competent he obviously performs a valuable service. There is no reason why a more general information service should not use or even incorporate such valuable sources of information. An experienced consultant would also be a valuable source of information about the needs of local industry. Similarly, where an industrial extension service exists the extension officers could well act as the agents of the information service, and form a valuable link with local industry. This could be particularly useful for the small-scale and informal sector.

The patent system

A potential component of a technical information service is the data which have to be supplied by those seeking patent protection, and which have to be deposited in a patent office, and by other owners of proprietary technology. This information on available patented technologies should certainly be used to the full, and can be a valuable source of information for any communications system. However, there are serious problems and limitations. The patent system as a whole, with its uncertain impact on developing countries, is now under scrutiny by UNCTAD[1] and the need for reform is generally accepted. Moreover, the coverage of patents is rather limited (in practice it is concentrated on chemicals, electronics, mechanical engineering and artificial fibres). Even more important, the knowledge disclosed in patents is only a small part of the total knowledge involved, and is usually deliberately limited in this way. However, this latter limitation is not really an obstacle to the use of patent information as a guide to available sources of technology, as distinct from its use for local R and D or actual production in developing countries. Even if much more information were disclosed in patents, the necessary skills and experience needed for actual production

---

[1] See UNCTAD: *The role of the patent system in the transfer of technology to developing countries* (New York, United Nations, 1975, Sales No.: E.75.II.D.6). See also S. J. Patel: "The patent system and the Third World", in *World Development,* Sep. 1974.

would often still have to be obtained by transfer of technology, and might only be available in a direct investment package. In the circumstances, patents mainly serve such functions as deterring smaller innovators with threats of expensive litigation or strengthening the oligopolistic control of markets against other technologically powerful rivals.[1] Neither of these two purposes would be generally beneficial for developing countries. The majority of the patents taken out in the developing countries are held by foreigners and remain unused.

Countries which have the technological power to supply the missing non-patented knowledge (Japan is often quoted as an historical example) can make very valuable use of patent information. Such countries would also have the capacity to develop a local process or product which is just outside the scope of the patented process or product, thus avoiding legal complications. This may apply to some of the technologically more advanced developing countries today.

The role of the patent system in the transfer of technology must of course be distinguished from its role in encouraging local innovation in developing countries. Protection by patents may be very valuable for this latter purpose, although possibly other rewards could be offered to innovators. At the present time, only about one-sixth of all patents granted in developing countries —themselves but a small fraction of the world total—go to nationals of the country. However, in future it may be possible to induce foreign patent holders (especially multinational enterprises) to develop local technologies and to prevail upon them to undertake local R and D work on the basis of patent protection.

An effective patent law and patent office can make an important contribution to the transfer of technology, if the drawbacks and abuses are properly controlled. For instance, the use of patents merely as a blocking device could be prevented by compulsory local leasing if the patent is not exploited. The effectiveness of the patent office also depends on good organisation and on the availability of sufficient resources to provide ready access to patented technology. Computerisation has obvious advantages, but may be beyond the resources of many developing countries. The cost could, however, be reduced by linking the patent office with the wider communications system discussed earlier and using common services, including computers. Technical and financial assistance could also be usefully directed to help with improving the organisation of patent offices.

At the World Employment Conference the Group of 77 and the Workers' group supported the suggestion that the Paris Convention of 1883 on industrial property—the basis of the present patent system—should be "drastically revised".[2]

---

[1] See Sanjaya Lall: "The patent system and the transfer of technology to less-developed countries", in *Journal of World Trade Law* (Twickenham, Vincent Press), Jan./Feb. 1976.

[2] Programme of Action, para. 62.

Channels of transfer

The multiplicity of sources and users of transferred technology has already been noted. There is an equal multiplicity of channels of transfer.[1] In fact, a good deal of technological transfer takes place in an informal way outside any institutional arrangements. This is especially true of the exchange of information between a parent company and its branches and subsidiaries, between larger firms and smaller firms acting as their subcontractors, between firms tied by linkages of supply and processing, and so on. This is paralleled in the development of new technology, much of which takes place outside the R and D system, or any other institutional system, as a result of spontaneous innovation at the workshop level in the form of learning, experience and the development of new skills. Machine suppliers also provide technology and information to their customers, often at no cost: but this is clearly a case of aid tied to the use of a particular machine or set of machines which may not represent the most appropriate available technology.

It must also be remembered that technology is embodied not only in machines but also in people. Skilled persons leaving one enterprise and moving to another carry with them the technical knowledge and information they have gained in their old jobs. The "interaction of people attending meetings, seminars and training courses" has been emphasised.[1] Such "embodied" transfer is often perverse—that is, it moves in a direction opposite to the one in which technological transfer is conventionally assumed to operate. With the "brain drain", there is a flow of technology from developing to developed countries (and the same is true when the subsidiary or branch of a multinational enterprise passes back its technical experience to the head office). Similarly, when craftsmen or skilled informal sector producers obtain posts in larger modern sector firms, there is an embodied flow of technology from the informal to the modern sector. The opposite flow occurs when a skilled worker from a modern sector plant leaves or retires to set up as an independent producer in his home area or in the informal sector.

Radio and television programmes, newspapers, technical periodicals, meetings of professional and technical associations and travel by consultants have all been mentioned as informal channels of transfer. It has been recommended that governments could do more to encourage such transfers, and that this might be more profitable than many of the direct actions now taken.[2]

---

[1] A full list of such channels in printed and non-printed forms is given in the *Report of the Expert Group on Information Exchange and Transfer of Technology (Vienna, 12-16 Apr. 1976)* (New York, United Nations Office for Science and Technology; doc. IAFTIS/XG.1/1), para. 17.

[2] ILO/UNDP: *Policies and programmes of action to encourage the use of technologies appropriate to Asian conditions and priority needs,* Report of the ILO Technical Meeting on Adaptation of Technology to Suit Special Market Conditions of Developing Countries (Bangkok, 3-14 November 1975) (Geneva, doc. ILO/TMAT/75/R.2; mimeographed).

Institutions involved in the transfer and dissemination of technology include applied or industrial research institutes. The coverage, size, resources and effectiveness of these vary widely. Ideally, the range of services they offer should include a team of experts and the capacity to test technical knowledge, including the capacity to set up pilot plants, before it is transferred. Their channels of transfer would include a technical reference library in their field, journals and bulletins available to interested producers, the organisation of exhibitions, meetings, seminars, and so forth. Many research institutions lack the resources for such a comprehensive and effective service. The strengthening of such institutions would clearly be an important part of any policy designed to increase the technological capacity of developing countries. The absence of suitable national infrastructure has been singled out as the "main reason" for lack of information. [1]

As well as research institutions, the scientific and technological departments of universities in developing countries, together with polytechnics and technical colleges, could serve as media for the transfer of technology. Valuable feedbacks for their own major tasks of training and research might well result if the institutions of higher learning were to play a greater role in the transfer of technology.

Apart from lack of resources, there may be other obstacles of a broader nature to the free and effective flow of technical information to those who need it in developing countries. The tradition of free exchange of information may be lacking. The relationship between government agencies and decision makers in the private sector may be one of mutual suspicion rather than of trust, confidence and co-operation. The same may be true of relationships within the government and the private sector. As far as relationships within the government sector are concerned, a recent meeting of experts from Asian labour ministries [2] suggested that governments should establish a national organisation charged with promoting the use of appropriate technology throughout the decision-making apparatus of the public sector. This organisation, it is proposed, could have its representatives in each ministry, who would assess every opportunity in the ministry for introducing appropriate technology at each decision level. It could also encourage small entrepreneurs to manufacture the simple capital goods required for appropriate technology. The skills which have been mentioned already as necessary both for the selection and supply of information and for its absorption and use, may be lacking. A certain climate of mutual confidence is necessary before communication systems can be fully effective.

Some support for the emphasis given to various factors in the previous discussion can be found in a recent study undertaken by the Institute of Science and Technology of the Republic of Korea. The Institute analysed

---

[1] *Report of the Expert Group on Information Exchange . . .,* op. cit., para. 18.
[2] Meeting held in Jakarta, 22-24 Mar. 1976.

the significant factors affecting the transfer of technology in the Republic of Korea to local industrial clients for whom the Institute undertook research under contract. This brought to light the following three key factors: *(a)* the technical soundness or success of the solution or advice; *(b)* the degree of support by top and middle management; and *(c)* the effectiveness of communication.[1] Of these three most important factors, the first refers to the quality and resources of the source of supply of information, the second to the attitude and co-operation of the recipient of the information, and the third to the nature of the communication system established between them.

Other important factors were the availability of qualified technical personnel (in many developing countries this will be more of a problem than in the Republic of Korea), the degree of complexity of the technology, and the marketing ability of the user. These factors refer more to the ability of the user to adopt appropriate technology when it is made known to him, rather than to the effectiveness of the communication system.

There seems no doubt that in nearly all developing countries there is a wide gap between the potential functions of a communication system as described above and what is in fact available. The task of bringing available technology to the attention of, in particular, the medium- and small-scale producers is greatly neglected in national technology policy; and even where its importance is recognised, the resources and skills required are not available. There is scope here for intensified national and supporting international action where the benefits, in terms of the promotion of development and the satisfaction of basic needs, could be very high. The same applies to the industrial extension service, of which the potential role in a communication system deserves strong emphasis.

Apart from the problem of the over-all insufficiency of information on available technologies, there is the additional matter of unequal access to whatever information there is. Scientists and technologists are often well served through documentation obtained from international or commercial sources. Firms connected with multinational enterprises or working as their subcontractors or suppliers can count on receiving the information they need from the multinationals. Larger firms with trained engineers and managers on their staff will also often be aware of the value of technical information and will often know how to obtain it. It is in the case of medium- and small-scale producers that the "communication gaps" are most pronounced.[2] Once again, this unequal access weights the scale against labour-intensive technologies, since labour intensity is inversely related to scale of production. It is in relation to smaller producers that an industrial extension service, as a link between the communication system and the user, would be particularly valuable.

---

[1] ILO/UNDP: *Policies and programmes of action . . .,* op. cit., p. 15.

[2] The Expert Group on Information Exchange and Transfer of Technology has described the technologies developed by, and suitable for, small- and medium-scale producers as "hidden technology".

Although a simplistic and uncritical transfer of technology from industrialised countries may introduce many inappropriate technologies into developing countries, among the mass of new technological developments there are nevertheless bound to be at least some which are capital-saving or are important and appropriate for developing countries for other reasons. This would represent a vastly greater reservoir of potentially useful information than the minute amount of technical research in industrialised countries that is undertaken specifically for the benefit of developing countries. [1] But whether the information potentially useful to developing countries arises from work undertaken specifically on their behalf or not, in either case there is the important problem of the "communications gap" which has to be resolved.

For the time being, only the technical information services of the major industrialised countries, or an international centre, would be capable of surveying existing and new technologies systematically and on a continuing basis, with a view to communicating to developing countries any technologies which seem to be valuable for them. An internationally agreed effort in this direction, perhaps with an agreed division of labour among the industrialised countries, with international support, would be an action well worth considering.

INTERNATIONAL CO-OPERATION IN TRANSFER OF TECHNOLOGY

There is a fairly rapid expansion of regional co-operation in developing countries, with a view to achieving the "unpackaging" of technology and the control of transfer pricing, and in other ways to obtain better terms in connection with technology that is "packaged" with direct investment, and generally to improve the bargaining position of the developing countries through the provision of better information, a better appreciation of the problems involved and the avoidance of competition with each other. The best-known and most effective case of such regional co-operation is the Andean Pact in Latin America. Such regional efforts can be supported and complemented by inter-regional and global arrangements. The Report of the Director-General of the ILO to the World Employment Conference [2] lists four ways of doing this, some of them already under consideration by UNCTAD, namely:

---

[1] The latter source would, however, become much more important if the targets of the United Nations World Plan of Action for the Application of Science and Technology to Development to increase the share of such research from 1 per cent to 5 per cent of the R and D expenditure of rich countries are accepted, or if the more general recommendations in the same direction of the resolution on development and international economic co-operation adopted at the Seventh Special Session of the United Nations General Assembly are implemented.

[2] ILO: *Employment, growth and basic needs...*, op. cit., p. 155.

(a) creation of small, specialised data-collection and analysis units to provide both (i) information on existing technology-transfer contracts and their terms, and on national regulatory systems, and (ii) analysis directed to identifying strengths and weaknesses of existing contracts and institutions, and possible directions and procedures for national action;

(b) provision of more effective technical assistance to countries seeking to develop institutional frameworks, train the personnel to man them and carry out important negotiations; this function might be combined with data collection and analysis or undertaken regionally by pooling national resources, as in the operations of the Andean Pact Secretariat;

(c) joint purchase of technology-use rights for several firms in several developing countries (either on a cost-sharing basis or by a single initial purchase followed by sublicensing); this might reduce transfer costs significantly in some cases;

(d) revision of the national and international patent system to make it more responsive to the needs of developing countries and to the attainment of the aims and objectives of a basic-needs strategy.

A number of information systems and services already exist within the United Nations and could form the basis of a more integrated international network. Such elements include the UNIDO Clearing House for Industrial Services, the various FAO agricultural information services, the ILO Integrated Scientific Information System (ISIS), and the corresponding information services operated by the International Atomic Energy Agency and the United Nations Environment Programme (International Referral Service). The United Nations General Assembly has supported the development of an industrial technological information bank and the establishment of technology transfer centres. This is an area where many initiatives are now taken at the national, regional and international level, and in the interest of efficient networking a co-ordinated approach under United Nations auspices is essential. As has been pointed out by the Expert Group on Information Exchange and Transfer of Technology, such a co-ordinated approach presupposes an over-all policy and agreed guidelines. [1]

The United Nations Inter-Agency Task Force on Information Exchange and the Transfer of Technology is working towards the establishment of a network for the exchange of technological information, and has developed detailed principles and recommendations for such a network. [2] At its second session in May 1976 it recommended that "organisations of the United Nations system and other organisations having substantive responsibility in the field of technological information and transfer of technology should develop their relevant activities as components of the over-all network, and

---

[1] *Report of the Expert Group on Information Exchange...*, op. cit., para. 43.

[2] United Nations, Economic and Social Council: *Science and technology: The establishment of a network for the exchange of technological information: Report of the Secretary-General* (New York, doc. E/5839, 14 June 1976; mimeographed).

in mutual co-operation make available their own information bases and information handling capabilities as appropriate".[1] Its report emphasises that "the bulk of the traditional sector, especially the urban and the rural poor, have been inadequately served because of the lack of methods for the dissemination of appropriate information among them", and calls for "special measures" to incorporate their needs in the network.[2] Special measures are also needed for proprietary technology.[3] As for the specific United Nations programmes and services, the need for compatibility between the various systems and for the harmonisation of programmes is stressed, as well as the need to help to improve capacity for technological information handling at the national and regional levels.[4] The latter task can be greatly facilitated if clearly defined national focal points are designated in all the developing countries.[5] This is clearly an area now ripe for constructive international action.

One of the main purposes of the proposed international network would be to remove the barriers to the dissemination of information on technological alternatives. The Integrated Scientific Information Service developed by the ILO—a computerised documentation system—could make a useful contribution to the over-all network. The World Employment Conference did indeed call upon the ILO to "strengthen its activities in the field of the collection and dissemination of information on appropriate technologies".[6] A beginning has already been made in developing inventories of alternative technologies: technical memoranda on technologies in civil construction are at present being prepared jointly by the ILO and the World Bank, and a proposal has been made for the preparation of similar memoranda in collaboration with UNIDO, on products and industries that are important to a basic-needs strategy.

---

[1] United Nations, Economic and Social Council: *Science and technology: The establishment of a network...,* op. cit., para. 46.

[2] ibid., para. 49.

[3] ibid., para. 50.

[4] ibid., paras. 62 and 65.

[5] ibid., para. 71.

[6] Programme of Action, para. 59.

ADAPTATION

4

ADAPTATION AND UPGRADING

Adaptation is an intermediate technology in the special sense that it is intermediate between copying or repeating the existing technologies (whether transferred from industrialised countries or already available in the country) and using a different, autonomously developed technology. It is a mixture of existing technology and new or existing scientific knowledge applied to the modification of available technology in the light of local circumstances. Adaptation is necessary or desirable in many dimensions: nature and quality of product, capital intensity, intensity with regard to other factors in the light of their relative scarcity and abundance, scale of production, and so on. To take the first example (nature and quality of product): for developing countries following a basic-needs strategy, production for the national market requires end products made to lower, or at any rate different, specifications, with simpler packaging, fewer models, fewer model changes and fewer unnecessary characteristics generally; the end product will perhaps be longer-lasting, require less maintenance, and so on. Corresponding changes are indicated in the raw materials, components and capital goods required, although the direction of change is not so clearly marked as in the case of the end product. For example, capital goods would not necessarily be made to lower specifications: shortages of skill and problems of repair and maintenance might well lead to a situation in which the appropriate capital goods would be made to higher, or at any rate different, specifications.

Closely related to adaptation, although theoretically distinguishable, is the improvement or upgrading of existing technologies. In both cases, an existing technology is the basis to which improvements and modifications are applied. In the case of existing technologies which may already have been adapted to the special conditions of the country, the problem may be more one of a general rise in efficiency than one of specific adjustment. However, the type of skill and the policies required for the adaptation and upgrading of existing technologies are very similar, and both are easier to achieve than the creation of entirely new technology. For this reason, the adaptation and upgrading of existing technology form the bulk of the national technological contribution in many developing countries.

Special attention in upgrading deserves to be given to the traditional and informal sectors. The technology used in this sector is "appropriate" in the sense that it has been developed for the specific markets and needs of the developing countries (except that, as pointed out before, in so far as this sector is the victim of unequal access to capital and other scarce resources its technology is *too* labour-intensive—more so than is necessitated by the over-all factor availability in the country). But it is often not efficient, and most of the producers in this sector fail to obtain incomes which enable them to satisfy their basic needs. Hence the upgrading of the technology used in this sector deserves high priority within a basic-needs strategy. The experience of countries following such a strategy (for instance China and Tanzania) suggests that the improvement of the simpler village technologies is the only viable approach to the gradual modernisation of a rural subsistence economy. The research conducted under the World Employment Programme and else-where, as well as the ILO comprehensive employment strategy missions, have led to a much greater concern with the problems of the informal sector. They have also resulted in numerous recommendations as to how the technology of this sector could be improved. One set of such recommenda-tions concerns the linkages with the modern sector through subcontracting, shifting of government orders and other means. Another set concerns a change in government policy from neglect and harassment to support and promotion. Of special interest in the present context is a series of recom-mendations to give direct technical support through advisory services, extension services, and research on and production of capital goods suitable for the informal sector. The analysis and recommendations of the mission to Kenya in particular have attracted much attention[1] and have evoked a very favourable response from the Kenyan Government.[2] A number of recent specific World Employment Programme studies have considerably added to our knowledge of the technological problems of the informal sector.[3]

At the same time, there still remains a lot to be learned. For example, not enough is yet known on how entrepreneurs and workers in the informal sector acquire their skills, to what extent there exist technological linkages between the formal and the informal sectors, and how one can encourage the transfer of technology from the formal sector to the informal sector, as well as the diffusion of such technology and skills within the informal sector. Indeed, the rapid industrialisation of Japan was made possible, to a great extent, by close collaboration between the two sectors, with larger enterprises in the formal sector encouraging and helping small enterprises to upgrade

---

[1] See ILO: *Employment, incomes and equality . . .,* op. cit., Ch. 13 and technical paper 22.

[2] Government of Kenya: *Sessional paper on employment* (Nairobi, 1973).

[3] All these studies are described and summarised in ILO: *World Employment Programme: Research in retrospect and prospect,* op. cit.

and adapt their technology. In some developing countries governments have been trying to encourage similar collaboration, but with little success. What factors determine success or failure? These are some of the matters which are receiving further attention under the World Employment Programme.

Some of the problems of creating incentives for adaptation through factor prices, income distribution and market structures were discussed earlier. But incentives are not sufficient; the capacity to adapt must also be there. This results in a vicious circle: in the least developed countries, where the "suitability gap" and hence the need for adaptation is greatest, the national producers' capacity to adapt is most constricted. Hence where the need is greatest, capacity is least, and vice versa. This vicious circle suggests the need for special international action regarding adaptation in favour of the least developed countries.

In most developing countries R and D institutions have the dual task of helping to adapt and upgrade known technologies and of helping to create new ones. The former task, even more than the latter, calls for close links with actual production, and these have been found to be missing in many cases. This problem of defective linkages will be discussed in more detail in the next chapter. Here it is relevant to note that any linkage defects on the part of R and D and related institutions will affect their effectiveness in the area of adaptation and upgrading even more severely than in the creation of new technology.

CORE VERSUS PERIPHERAL PROCESSES

In adaptation, research has confirmed the importance of distinguishing between the core process and peripheral processes. In general, adaptation is easier and more widespread for the peripheral activities which relate to the flow of inputs and outputs on either side of the basic machine process, to and from the machine. While research under the World Employment Programme and elsewhere has confirmed the broad truth of this generalisation, it has also brought to light a number of complications and exceptions. Generally, the peripheral activities are more akin to the situation in construction, transport and services where the labour-intensive potential is greater than in industry, as represented by the core process. In the light of recent research, however, it would be wrong to be "core-pessimistic" in regard to the adaptation potential of the basic machine process. The greater ease of adaptation in the peripheral activities may be partly a matter of limited capacity to adapt: peripheral activities offer easier and more obvious opportunities for labour-intensive adaptation. If this view is correct, an enhanced technological capacity in the future in the developing countries would result in a more even spread of adaptation between core and peripheral activities. The difference may be partly due to the fact that the core process is determined rather by engineers who may be prejudiced against any deviations from the latest technology, whereas the peripheral activities are determined rather by commercial managers who pay more attention to factor prices and local markets. To the extent that this explanation applies, a revised method of training engineers and increased collaboration between engineers and commercial managers may gradually reduce the gap in adaptation between core and periphery.

With regard to Asia, it has been pointed out that adaptation of the machine process is often stimulated by the "indigenisation" of machinery.[1] By this is meant the copying and local reproduction of an imported machine, subject to the observance of patent regulations. Apparently, considerable cost advantages have been achieved where capital equipment has been indigenised and has been adapted in the process, rather than continuing to be imported. India, the Republic of Korea and the Philippines are quoted as evidence that copying, through a better understanding of the machinery, leads to adaptation.[2] As was pointed out in Chapter 3, copying also avoids repeated and cumulative payments for technical know-how. The great historical example for this is Japan.

Research has indicated that there is a two-way relationship between the local reproduction of machinery and the adaptation of technology. Local reproduction can lead to adaptation, but adaptation can also lead to local manufacture and thus help in building up a national capital goods industry— partly because the adapted machinery may be easier to produce domestically, and partly because it may not be available from abroad on satisfactory terms.

PRODUCT ADAPTATION

Product adaptation has already been mentioned as an important aspect of adaptation. It is given special significance by any movement towards a basic-needs strategy. Both "utility" versions of modern sector goods and improved and upgraded versions of traditional and informal sector goods play a part in product adaptation; in addition, there are entirely new products which a consistent basic-needs strategy combined with strengthened mass markets and an increased technological capacity of developing countries is bound to bring forth. To take the first case of "utility" version of modern goods, models of various "Asian cars" may be mentioned as a remarkable example. A number of foreign car-manufacturers established in the Philippines have introduced such models in order to increase the local content of production without raising costs or sacrificing quality to any appreciable extent. An extreme case is provided by one manufacturer in whose car bodies bamboo and wood, which are in plentiful supply in the Philippines, are used. Moreover, work on bamboo and wood, rather than metal, calls for simpler skills.

---

[1] See ILO/UNDP: *Policies and programmes of action . . .*, op. cit., p. 18.
[2] ibid.

SOURCES OF ADAPTATION

The adaptation of technology requires engineering design skills and has obvious training implications. On the strength of examples drawn from China, it has been stated that the vigorous development of engineering design skills can to some extent replace and bypass the conventional R and D machinery of laboratories, pilot plants, and so on. It has consequently been recommended that relatively more emphasis should be given to design skills and less to conventional R and D. [1] One could add that, if R and D institutions are to some extent relieved of their tasks of adaptation, they would then be able to concentrate on their really major task: the creation of new technologies where research laboratories and pilot plants are indispensable.

The major source of adaptation should normally be at the plant level. It was not until comparatively late in the development of the now industrialised countries that scientifically oriented R and D institutions not associated with individual firms evolved on any significant scale. Up to that point technology had developed out of shop floor experience and rudimentary research, first in small and then in gradually larger specialised departments of the company. This type of development, within companies rather than through outside institutions, ensured that there was a close link with concrete problems and tended to place emphasis on the continuing adaptation and introduction of technology into the actual production process.

In the earlier economic history of the industrialised countries, the technology at any given moment was much more in harmony with the factor endowment than is the case with the technology at present being transferred from industrialised to developing countries. Hence the task of adaptation is much greater and more urgent in the developing countries of today, and it does not follow that the simple company-based machinery would be sufficient. Moreover, in countries where much modern sector production is controlled by external decision makers such as multinational enterprises, and where at the same time efforts are made to change the income distribution and move to a basic-needs strategy, it would seem less advisable to leave everything entirely to intra-company arrangements. This would also leave out the traditional, small agricultural and informal sector and other non-company modes of production. However, in spite of such qualifications, a vigorous expansion of capacity for adaptation at the plant level (as in the case of the development of engineering design capacity) will tend to reduce the load of the institutional structure, which is bound to suffer from constraints and insufficient resources of all kinds.

ADAPTATION AND RURAL DEVELOPMENT

The agricultural sector occupies a special place as regards the adaptation of technology, in that the need for adaptation to special local conditions

---

[1] See ILO/UNDP: *Policies and programmes of action...*, op. cit.

and resources is much clearer than in the case of industry, and was recognised and acted upon much earlier. The new technologies under the Green Revolution, based on high-yielding varieties of seed, have provided another, more recent, example of intense efforts to adapt to specific local traditions. Moreover, agriculture could not be served to the same degree by informal intra-company arrangements. The necessity for more public action and more institutionalised arrangements was therefore recognised right from the beginning. These institutions have also been better able to keep in touch with direct production problems. [1]

Rural non-agricultural technology also deserves special government action. Like agriculture, the units of production in rural and cottage industries are small, so that the role that intra-company arrangements could play with regard to the adaptation of machinery would be somewhat limited. Rural development has emerged as a high priority, both because the majority of those below the basic-needs standards in most developing countries live in the rural areas and because rural development will reduce the problem of urban unemployment and poverty by reducing rural-urban migration. Moreover, with rural and cottage industries one would be faced with a normally appropriate labour-intensive technology, so that the task of adaptation is largely one of upgrading and improvement. The rural and informal sectors have also been shown to be important potential sources of indigenous innovation and efficiency.

One should not, however, neglect the possibility that important linkages might be created between the urban formal sector and the rural informal sector. It is fairly well known that, in the earlier stage of the industrialisation and postwar recovery of the Japanese economy, such linkages played a crucial role, particularly in export-oriented garment and other light industries. A more recent example of no less significance is provided by *shibori* (silk fabric) processing in the Republic of Korea, where trading firms act as intermediaries for marketing and the transfer of technology and skills to the rural sector and where the rural co-operatives act as agents for the diffusion of technology and skills within the rural sector. [2]

The experience of Tanzania, which was examined in detail under an ILO/UNDP project concerned with appropriate technology for rural development, is also interesting. [3] It shows that quite often too much attention is paid to "modern" high-yielding biological technology and its associated

---

[1] This has by no means prevented the new technology of the Green Revolution from giving rise to a number of serious social and economic problems. These, however, were problems of wrong economic and social structures and a "communications gap" with small farmers, and thus beyond the scope of the purely technical adaptation of machinery, which cannot be held responsible for the failure to prevent the undesirable social and economic consequences of the new technology.

[2] For a brief discussion of this industry, see S. Watanabe: "International subcontracting, employment and skill promotion", in *International Labour Review,* May 1972, pp. 433-434.

[3] For details, see George Macpherson and Dudley Jackson: "Village technology for rural development: Agricultural innovation in Tanzania", ibid., Feb. 1975.

inputs (fertilisers, pesticides, irrigation, and so on), to the neglect of farmers' tools and implements. As part of the new policies for rural development in Tanzania, the Tanzanian Second Five-Year Plan therefore proposed that the Tanzania Agricultural Machinery Testing Unit (TAMTU) would design "simple and inexpensive farm implements which could easily be made from available raw materials". Thus the objectives of an appropriate technology policy should be to utilise local resources and skills for the design and development of technology which is more productive than the traditional technology yet which is within the reach of the farmers and other poor groups. Self-reliance at the local level implies that technology be adapted to the needs of the local community and to their resource position, and that their active participation and involvement be invited.

Rural development also clearly calls for technological and organisational changes which will make the task of the adaptation and upgrading of technology easier: these include the formation of village co-operatives or groups of co-operatives, the development of efficient techniques of low-cost road construction based on local materials combined with the use of volunteer and temporary local labour, and similar forms of labour mobilisation for small-scale irrigation, terracing, drainage and other land improvement projects.

Improvements in the domestic water supply would be especially important, not only for improving health but also for reducing the heavy workload of rural women in fetching and carrying water over long distances. The same applies to any improvement in energy and power supplies which would reduce the burden of collecting and carrying firewood, whilst at the same time yielding other great environmental benefits in preventing deforestation and soil erosion, or in setting free manure for fertilising the land. This would also help with food preservation, once again reducing the workload of rural women. The emphasis on rural women is justified by much recent research, which reveals the great benefits for the welfare of rural families that could result from an easing of the extraordinary pressures under which rural women usually have to live and work. In this area considerably more research and data are needed and should be given priority, in that it is a topic bearing directly on the essential needs of the masses of the poorer rural populations. Work in this field has already begun within the framework of the World Employment Programme.[1]

Apart from the adaptation and upgrading of existing facilities through the use of known technologies, the development of new technologies (discussed in the next chapter) is urgent, particularly so perhaps as regards the provision of a decentralised power supply based on solar energy, and the development of small generators (already attempted in several countries) and bio-gas processes.

---

[1] See, in particular: I. Palmer: "Rural women and the basic-needs approach to development", in *International Labour Review*, Jan.-Feb. 1977.

ADAPTATION AND SECOND-HAND EQUIPMENT

The use of second-hand equipment as a form of adaptation of technology has already been mentioned in Chapter 2. Research under the World Employment Programme has thrown much light on this question—not so much by showing whether the use of second-hand machinery is universally recommendable or not, but by clarifying the situations where it can be useful, the risks involved and the conditions which are required for its selective use. Some of the results (for instance, on the seriousness of the spare part problem) are contradictory; but this again may well reflect differences in the underlying conditions in the various situations examined. In studies of Brazil it has been found that second-hand machines are more suited to production for the local market than for export production—a plausible finding. In Japan, it is stated, "the relative share of second-hand machinery . . . still remains high for small factories, with gradual tapering off for large factories. It has been suggested that one of the reasons for the small enterprises keeping pace with large ones in Japan has been the effective utilisation of second-hand machinery." [1] On the other hand, the conditions in which import-substituting industrialisation has often been promoted have not been favourable to the use of second-hand machines, even where this would have been useful as a method of adaptation of technology.

One of the objectives of research conducted under the World Employment Programme is to increase knowledge of the complex technological factors involved in the use of second-hand machinery.[2] This also suggests that any conclusions have to be eclectic and can be reached only by a study of each particular situation. For instance, second-hand looms were found to be less productive than new ones, but this was not the case with spinning frames. This leads to the more general hypothesis that machinery which is initially simple and more robust may also lend itself more readily to second-hand use—a cumulative effect of appropriate and adapted technology. The research has brought out the multiple factors of risk and uncertainties, while the same and other studies have also discovered a potential for the selective use of second-hand machinery. The relative price of second-hand and new machinery is only one factor in a complex situation.

There is also conflicting evidence about the seriousness of the spare part problem which serves to emphasise the importance of the fluctuating efficiency of international trade in spare parts and the availability of specialised local firms dealing in spare parts. The spare part problem was also shown to be linked with the reasons why the second-hand machine was for sale in the first

---

[1] A. S. Bhalla: "Small industry, technology transfer and labour absorption", in OECD Development Centre: *Transfer of technology for small industries* (Paris, 1974), p. 111.

[2] Particularly the study by C. Cooper and R. Kaplinsky: "Second-hand equipment in developing countries: Jute processing machinery in Kenya", in ILO: *Technology and employment in industry,* op. cit.

place. If it is being sold because of real obsolescence, new machines of the same type will not be produced and thus it will be more difficult to obtain spare parts. The problem of repair and maintenance has been found to be less serious in cases where the required skills are locally available, because of the low wages in developing countries and the typically labour-intensive nature of repair and maintenance activities. The disconnection of the machine, and its transport and re-installation, have also been shown to affect different types of machinery differently.

The possibility has emerged from the studies that the use of old machinery in developing countries might actually contribute to the capacity to create indigenous machine shops and learning processes through repair, maintenance and the copying of older designs. But this will call for special action, usually governmental action. For instance, the ILO comprehensive employment strategy mission to Kenya recommended that consideration be given to the use of the large East African Railway workshops as well as small machine shops to give technical support to industries in the informal sector, which often use second-hand machinery. The studies show that the best policy is one of a discriminating use of second-hand machinery with a certain reticence, and that neither indiscriminate bans nor an indiscriminately open door for second-hand machinery are likely to be optimal policies. The organisation of trade in second-hand machinery, with an international classification of machine quality, has also emerged as a potential area for useful international action in a field that is at present neglected.

Second-hand machines may be especially useful where new machines of the same type are not available—in other words, where the old machine has been displaced by process innovations in the industrialised countries (that is, of course, if the process innovation is not appropriate for developing countries, either because of its technology or because it contradicts a basic-needs strategy). The obvious risk relates to the cost of installation, the higher running cost of production, the greater uncertainty, the problems of after-sales services, repair and maintenance, hazards for workers, and so forth. On the other hand, where repairs and maintenance skills are available this provides employment in a very labour-intensive activity also suitable for small-scale units at low social cost. The arguments for and against second-hand machinery are complex. The question is important enough to call for public policies in the matter. Better organisation and control of the market in second-hand machines could reduce the disadvantages and risks attached. An analysis of the reasons why the machine is being discarded in the industrialised country can often be helpful as a basis for a decision whether its use would be indicated. Advice to potential users on this and other matters could help them to make wise decisions and avoid mistakes. Such advice could be organised on a national, regional or international basis.

Reluctance on the part of buyers in developing countries to purchase used machinery results partly from lack of information, and much less so from reasons of prestige and pride. The Kenyan case study on jute processing machinery[1] examines the question of availability of second-hand equipment in this context. At present most of the used machinery is imported into the developing countries from advanced countries. The trade in used equipment is irregular and sporadic. The jute equipment market is peculiar in the sense that there are hardly any supplying centres except Dundee in Scotland. For the buyers in developing countries this means that they cannot shop around for different sources and have a very poor bargaining position. There is another disadvantage in the sense that the developing countries themselves hardly present any source of supply for used equipment because of the infancy or non-existence of indigenous capital goods industries.

GOVERNMENT ACTION

Quite apart from the special case of second-hand machinery, government action in assisting and inducing the adaptation of technology is needed. This is not limited to direct action. Indirect action which creates a framework of incentives for continuing adaptation may be as important as, or more important than, direct action. Indirect action could include policies of tax relief, depreciation allowances, control of imports, and so on, extended not only to promote new investment but also to cover cases of the adaptation and upgrading of existing technology. Direct measures would include technical and financial assistance, flexible quality standards, and so on. The adaptation of technology used by multinational enterprises may also be promoted by effective government negotiation at the time the admission of the enterprises is approved.

Government policy can also be important in specific areas of the adaptation of technology. For example, reference has already been made to government action which could be useful in ensuring the wise and discriminating use of second-hand machinery. Such action includes advisory services to potential users of second-hand machinery; the establishment of quality standards and quality classifications for second-hand machinery, preferably as part of an international effort; arrangements such as that recommended by the ILO comprehensive employment strategy mission to Kenya regarding the use of the large public sector East African Railways workshop (see above); government help in collecting and disseminating information on experiences with the use of second-hand machinery and the proper guarding of such machinery; assistance in organising the market for second-hand machinery on a less haphazard basis. All these are major tasks, but they may be justified by the important benefits of promoting good uses and preventing bad uses. Similar government action could be mapped out regarding other aspects of technology adaptation, as has been mentioned above in the case of rural technology.

---

[1] Cooper and Kaplinsky: "Second-hand equipment in developing countries...", op. cit.

In various situations, government import regulations have contributed to the adaptation of technology. For instance, from India it is reported that government restrictions on importing various oils led to the development of methods of making soap with local oils.[1] Generally, foreign exchange shortages with or without deliberate government measures will result in all types of technological change tending towards the greater use of local resources: a choice of technologies depending more on local and less on imported materials; the adaptation of technologies based on local materials; and the development of new technologies based on local resources not previously used. The effect of government policies is thus not limited to adaptation but affects technology across a broad front.

At the World Employment Conference, foreign firms were called on, in response to the national legislation of developing countries and in negotiation with them, to adapt technology to the needs of the host countries, and progressively to substitute national for imported technology.[2]

---

[1] Information supplied by Unilever and reported in Frances Stewart: "Technology and employment in LDCs", in E. O. Edwards (ed.): *Employment in developing nations* (New York and London, Columbia University Press, 1974).

[2] Programme of Action, para. 55*(b)*.

NEW TECHNOLOGIES

5

THE NATURE OF THE TASK

New technology is necessary for two major reasons. First, with the present unsatisfactory distribution of R and D and of actual production experience, many of the problems arising in and methods that would be appropriate to developing countries have not been taken into account in the selective expansion of science and technology. Impressive as this expansion has been when it is looked at from a global viewpoint and when the interests of the majority of mankind living in developing countries are considered, it is seen to have been on a narrow front and to have left many important areas beyond the present frontiers of knowledge. Second, new technology may also be needed by the developing countries because the appropriate technology, although within the present frontiers of knowledge, is in fact not available. It may not be available because it is outdated and because the present suppliers of technology from the industrialised countries have no direct interest or willingness to supply this technology, or goods embodying it, to the developing countries. A slightly different case belonging to the same category occurs where the appropriate technology is in fact available but only at such an excessive price or under such disadvantageous conditions (for instance, as part of an investment package) that it becomes rational[1] to create a national technology even though an acceptable imported technology is theoretically available.[2]

The best method of creating the required new technologies may differ somewhat in the two cases distinguished above. Where a developing country, because of its specific situation and requirements, is faced with a problem beyond the present frontiers of knowledge, it will often be necessary to

---

[1] From the point of view of the developing country. Global rationality may be better served by improving the conditions under which the required technology is made available to the developing country.

[2] For an elaboration of the statement made above, see UNCTAD: *Transfer of technology*, Report prepared for the Fourth Session of UNCTAD, Nairobi, May 1976 (Geneva, doc. TD/190, 31 Dec. 1975; mimeographed), and also, elaborating for the case of a specific industry, Bell et al.: *Industrial technology and employment opportunity . . .*, op. cit.

encourage new research, possibly even of a fundamental nature. Such research could also be carried out in the industrialised countries where a very much larger capacity for fundamental and basic R and D work exists. The United Nations World Plan of Action[1] has in fact identified a number of such priority areas and has suggested that, to a considerable extent, the necessary new research should be carried out more immediately in the industrialised countries by means of their diverting an agreed proportion of their R and D expenditures to these priority areas. This was considered necessary as a short- and medium-term solution and is based on the assumption that the results of the research would be readily and fully transmitted to the developing countries and would be tested and applied to the specific conditions of each country by continued development within the developing countries. As a long-term solution, it would of course perpetuate technological dependence. However, such a contribution by the industralised countries to the research needed for new technologies in high priority areas could be, under favourable conditions, a particularly effective form of aid—probably more effective than many forms of aid practised at present.

The Declaration of Principles and Programme of Action adopted by the World Employment Conference established the principle that the promotion of research should be "a fundamental priority"; that it should "mainly be undertaken within, and under the direction of, the developing countries themselves or in corresponding regional or subregional bodies where these exist, with the technical and financial assistance of international and other agencies"; and that it should "contribute towards the satisfaction of basic needs" (para. 53). The informal urban and rural sectors are specially singled out for accelerated appropriate technological advancement (para. 54).

The second case may not require a great deal of fundamental research since the technology to be applied exists in principle, although it may not be available in satisfactory form. Where the technology has been supplanted in the industrialised countries, material such as blueprints, copies of machines and technical literature may be available to form a sufficient basis for putting the emphasis on development work rather than on research work. Thus the balance of work between R and D, and within R between basic or fundamental R and applied R, will partly depend on the relative importance of the two situations which require and justify the creation of indigenous technology.

One of the big problems in creating new technology is to ensure that the new technology is in fact appropriate, that is, based on the direct requirements of the actual producer and in line with the development objectives of the government (for example, a basic-needs strategy). This is the communications or linkage problem which will shortly be discussed in more detail.

In the industrialised countries, and especially during their earlier industrial

---

[1] United Nations, Department of Economic and Social Affairs: *World Plan of Action for the Application of Science and Technology to Development,* op. cit.

history, the bulk of technological innovation arose directly from within the production plant on the basis of shop-floor needs and shop-floor experience. Technical innovation precedes the existence of formal R and D institutions by a considerable margin, whether in the public or in the private sector. The important study on can manufacture quoted above shows that the technology of can manufacture, which is now considered appropriate for the developing countries studied, was created by simple workers and small managers, "unfinished mechanical engineers". Even at a much later stage, the R and D establishments of private companies still consisted of the odd scientist and/or engineer working in a corner of the plant; public R and D was unknown, except perhaps in the field of agriculture. The assumption that the creation of new technology *must* come through R and D institutions is taken from the example given by the present industrialised countries and disregards past experience. For this reason, in the creation of new technology it is at least as vital to create the conditions, incentives and skills for innovations at the shop-floor level arising from direct production experience and to see that such innovations are disseminated and applied. This also is a question of the training and skill of those engaged in actual production, as distinct from the training of scientists, engineers and technologists suitable for work in R and D institutions.

"National" or "indigenous" R and D and resulting "national" technological capacity, to be also appropriate in the context of a basic-needs strategy, will have to give high priority to the needs of small farmers, rural and small-scale industries and the informal sector.[1] This high priority applies quite particularly to the provision of improved tools and equipment to enable producers in these priority sectors to raise their productivity and incomes on a permanent basis. This provision of material is also an important element in the creation of a national capital-goods industry as a dynamic force for future technical progress. Thus the objective of an appropriate national R and D system is not simply to reduce technological dependence and close the technological gap—important as such objectives are—but also to narrow the technological gap within the developing country. The distribution and allocation of R and D resources is therefore as important (always assuming that a basic-needs strategy has been accepted) as its total volume. This is an aspect which must not be overlooked when setting over-all quantitative targets for the total volume of R and D in developing countries. It should be emphasised that the corresponding proposals in the United Nations World Plan of Action were closely tied to suggested priority areas for the allocation of funds under the suggested R and D objectives.

The statement that the advance of technology—determined by the industrialised countries—has been on a selectively narrow front only, shows that the picture of a simple accumulation of knowledge, an increasing store of knowledge on which the latecomers can draw, is false. What we see is much

---

[1] This has been explicitly recognised in the Programme of Action, para. 54.

more comparable to a flow than to simple accumulation. New science and new technology are created at one end, but displace science and technology that existed previously. Being located in the rich countries with their dynamic search for new knowledge and their vastly and increasingly different requirements, priorities and factor endowments, it is not surprising that the knowledge displaced or submerged at one end may well be more useful to the developing countries than the new knowledge added at the other end. It is therefore by no means clear, from the viewpoint of the poorer countries, that there is in fact an accumulation of knowledge in the relevant sense.

ROLE OF LOCAL CAPITAL-GOODS INDUSTRIES

New technologies, whether new in the sense of being outside the present frontier of knowledge or new in the sense that they are no longer available in the market, specifically require a local capital-goods industry, in addition to design capacities and skills. The development of a local capital-goods industry provides a favourable basis for appropriate technological innovation. The existence of local capacity for the production of machinery will help to remove the lack of confidence in the national capacity for local technological innovation. There is considerable scope for the increased production of machines, as well as for the adaptation of imported machinery. The local production of machinery can often reduce the capital cost of machine use, since the importing of machinery includes a monopolistic or oligopolistic pricing element and is often subject to sales pressures. The machine-producing sector, contrary to a widespread belief, is also one of the most labour-intensive of all the manufacturing sectors. In addition to being itself labour-intensive, it can affect the labour intensity of the whole economy. Furthermore, it makes possible a particularly high level of training and skill development. Above all, local machine-producing capacity is an important element in technological independence.

There are thus many reasons for giving the local production of machinery very high priority indeed under a policy of spreading appropriate technology. New and appropriate technology often requires equipment which is no longer available in the market but for which blueprints exist and which are often well within the developing countries' current technological capacity, in addition to providing a good basis for dynamic future technological improvements. One major reason why in many developing countries the local production of machinery is relatively little developed derives from the nature of import-substituting industrialisation. Under this strategy, when industries at the end of the production line are established, imports of capital goods are usually freely admitted or even subsidised. This leaves little room for the development of the local production of machinery, especially when it is coupled (as it often is) with a transfer of technology based on the type of equipment developed in the industrialised countries.

The development of new technology based upon existing knowledge no longer used in the industrialised countries has been described as "do-it-yourself technology".[1] In this case the major part of the skills required is not so much in sophisticated research ability as in strong design and drawing offices and accumulated engineering expertise among production and design engineers. The role of the scientist and the research engineer will be correspondingly reduced.

The capacity for a do-it-yourself technology will also, perhaps paradoxically, strengthen the bargaining position of developing countries with machine suppliers, multinational enterprises and the like which wish to transfer an inappropriate or excessively costly technology. The existence of such national capacity may result in better terms for technology transfer, possibly so much so that the transfer of what was previously inappropriate technology now, on the better terms available, becomes appropriate. The paradox in this case is that, even though the do-it-yourself technology is not in fact used, the capacity to use it if necessary will itself prove valuable to the developing countries.

R and D for the development of new technologies can often be combined with development of local capital-goods production. A well known example is afforded by the International Rice Research Institute (IRRI) in the Philippines, which has developed not only new varieties of rice but also the complementary, low-cost farming implements appropriate for the cultivation of the new varieties by small farmers in Asia. It has recently embarked on an international subcontracting programme involving agencies, institutions and engineering firms in nine Asian countries. Under this programme, models from the IRRI design programmes are sent to the subcontractor's country and tested under local environmental and economic conditions. An effort is also made to interest local manufacturers in the designs. In addition to the subcontracting programme, information on particular designs has been furnished at the request of many independent manufacturers throughout the Asian region.[2]

R AND D INSTITUTIONS

In the development of new indigenous technology considerable emphasis is being placed on the role of research institutes in conducting R and D work. There is a surprising number of such institutes in the developing countries, and although the United Nations World Plan of Action has pointed out that the percentage of gross national product spent by the developing countries on R and D activities is much less than in the industrialised countries,[3] the

---

[1] Bell et al.: *Industrial technology and employment opportunity . . .,* op. cit.

[2] See ILO: *The poor in Asian development: An ILO programme,* Report of the Director-General, Eighth Asian Regional Conference, Colombo, 1975, p. 72.

[3] See Chapter 2.

absolute sums involved are quite considerable. It must also be taken into account that in the industrialised countries the bulk of R and D expenditure occurs in the private companies, so that public R and D institutions of various kinds are only a minor part of R and D, whereas in the developing countries the exact opposite is true. Hence, when we speak of expenditure on public research institutions, the disparity between developing and industrialised countries is by no means as large as the over-all figures suggest.

There seems to be general agreement that, in relation to the sums involved, the results have been disappointing in terms of the creation of more appropriate technology, particularly labour-using technologies and technologies directly geared towards benefiting the poorer sections of the population, specifically the small farmer and small producer in the informal sector. There are, of course, brilliant exceptions and success stories. The Green Revolution in rice originated in a research institution in a developing country, although not a governmental institute (the IRRI in Los Baños), and this was certainly a potential breakthrough in a vital area, even though in some cases the results were affected by a disregard of social and political factors. Similarly, the Central Leather Research Institute in India and the Institute for Industrial Research of the Republic of Korea are often quoted as examples of successful operations; and there are many others. Still, the judgement must stand that the over-all contribution of R and D institutes in developing countries has been disappointing. It is pertinent to consider the reasons for this, with a view to drawing policy conclusions for future improvement.

One reason emerges from a comparison with the earlier history of the industrial revolution in the European countries. The greater part of technical innovation and creation of what was then an overwhelmingly appropriate technology was not through R and D institutions but through shop floor innovation. As has been pointed out already, even the company R and D departments originated in this manner. This method of technological innovation from the ground upwards had the great advantage of ensuring that any research and development was guided by shop floor problems and shop floor requirements. As it has been said, it was industry which guided science and technology, not the other way round. At that time, this was the right and effective method. The hypothesis may be advanced that an attempt to institutionalise innovation prematurely in the earlier stages of the industrial revolution would have retarded progress rather than advanced it. This is not to say that, in the entirely different conditions of today, the industrialised countries are not justified now in giving much more emphasis to institutional R and D for the creation of the much more sophisticated new technology required.

In the developing countries of today, one factor which has emerged from all the studies as a major reason for the disappointing contribution of R and D institutions to the creation of appropriate new technology is the lack of contact with shop floor problems and shop floor experience. The fault presumably lies on both sides: lack of contact of the scientists and technologists in the R and D institutions with the actual producers at the plant level, and a failure of producers at the plant level to bring their problems to the R and D institutions and take advantage of the contribution which the latter could make.

Several possible solutions have been proposed to remedy the aloofness of scientists and technologists in R and D institutions regarding the needs of actual producers. One obvious solution would be to send the scientists and technologists frequently into production units for first-hand observation of problems at the workshop level; apparently this is being done in China on a considerable scale. Another solution would be to organise the R and D institutions so that they work only or mainly on contract from actual firms; this apparently is done in the Republic of Korea. The assumption is that, when work is done on contract and has to be paid for by industry, it will reflect the real interests of the market and maximise the chances for the actual utilisation of the new technology; at the same time, the payments received under contract will help to finance the R and D institutions. The trouble with this solution is that it would not apply to those most in need of new technologies, that is, the small-scale and informal sector producers. It would limit access to the R and D institutions to those with the greatest capacity to pay for the services, excluding those most in need of these services in the context of a basic-needs strategy.

To ensure that producers bring their problems and experiences to the R and D institutions is partly a question of mutual confidence between producers and public sector institutions and partly a question of a lively interest in shop floor innovations, including a willingness to pursue such innovations and to search at the R and D level for solutions which cannot be found at the shop floor level.

In the absence of sufficiently close contact with production units, the scientists and technologists in the R and D institutions are apt to give priority to pure or basic research rather than to applied research. They are also apt to take up the problems discussed in the scientific literature; but these are largely the problems of the industrialised countries, and consequently the R and D institutions of the developing countries may make marginal additions to the R and D work of the industrialised countries. All the empirical studies confirm this danger.[1] In many cases, the results of R and D work are lying on the shelf, or perhaps they have resulted in an article in one of the scientific or technical journals usually published in an industralised country. This is a phenomenon which has already been discussed as the "internal brain drain",[2] which will often lead to an actual brain drain.

A third reason is the inability, for lack of contacts, to take the R and D work to the stage of production through pilot projects, demonstration projects

---

[1] See, for example, Frances Stewart: "Technology and employment in LDCs", in *World Development*, Mar. 1974.

[2] See Chapter 2.

and help with "upscaling" at the stage of actual introduction into the production process. Of course, pilot projects and, particularly, demonstration projects are costly—the D tends to be more costly than the R—although in many cases demonstration plants might pay their own way. Pilot and demonstration plants, therefore, are perhaps particularly suitable for support from aid organisations or through international assistance. It may be recalled that when the United Nations Development Programme (UNDP) was initially set up to deal with "preinvestment" support, pilot plants were specifically included within its competence, although in practice little was done in this area.

The fourth approach is to associate the R and D institutions from the very beginning with technical advisory services or provision for answering technical questions which will ensure contact with productive units. A strong advisory board composed of persons actually responsible for production decisions would also be useful in this connection.

The organisation and spirit of R and D institutions is also very important for the success of their work. The successful innovation and creation of new technology depends very much on the presence of highly skilled and motivated personalities. A successful and well organised institution will give full scope and support to such personalities in their work, but a bureaucratic or badly organised institution can easily frustrate or deter them.

As will be discussed in more detail,[1] the proper functioning of R and D institutions also largely depends on a careful definition of their functions within the framework of an over-all science and technology policy. This in turn must be based upon development priorities and objectives, such as the implementation of a basic-needs strategy. Such a technology policy must also encourage maximum innovation and adaptation within the production units themselves. This is important so that the R and D institutions may specialise in tasks which can be better carried out at the R and D level than at the production level. This will be particularly the case where technical processes are of special interest to the developing country but not to developed countries, so that no previous stock of knowledge exists on which to draw. Maximum innovation at the production level is also useful for creating a flow of requests and problems from the production level to the R and D institutions, thus serving to prevent their isolation. Innovation at the production level depends partly on incentives and attitudes and partly on levels of skill and training. Hence training programmes and the development of required skills also form an indispensable part of a science and technology plan integrated with the general development plan.

One factor which may limit the effectiveness of R and D institutions is a lack of confidence placed in them at the production level, so that the import of foreign technology is preferred and is assumed to enhance the acceptability of the product. This has been described by UNCTAD as an "asymmetry of

---

[1] See Chapter 7.

70

initiative".[1] The boldness needed to challenge the received technology is absent. Such a lack of boldness or asymmetry of initiative has clearly emerged from the previously mentioned study of can manufacture in developing countries.[2] It is not difficult to see that this situation sets up a vicious circle in that the domestic R and D establishments are bypassed, and in response they withdraw into theoretical and academic work. The study of can manufacture also provided a good illustration of the frequently occurring situation in which such a lack of boldness is unjustified and is due to a considerable extent to a failure to realise that what is needed is not an entirely new technology, but rather the reproduction of a technology which is part of the existing stock of knowledge, although it is not or no longer available through imports on anything like acceptable terms.

The problem of contact with smaller firms and informal sector producers which have no R and D facilities at the plant level and where the difficulty of establishing contact between national R and D institutions and the production level may be greatest can, to some extent, be solved by forming research associations of small- or medium-scale enterprises. Such research associations can at the same time stimulate innovation close to the production level and serve as a link between the production level and R and D institutions at the national level. However, the instrument of a research association is hardly adequate to reach the informal sector.

It has often been pointed out that the risk of aloofness by scientists and technologists as regards actual production problems is due to the special characteristics of their education and training. This education and training is often based on a model directly imported from industrialised countries, and the technology implicitly or explicitly assumed in their training is different from that in developing countries. Moreover, at the higher levels, many scientists and technologists receive their advanced training in industrialised countries so that if, in the course of their training, they have contact with the actual users of technology, it will again be with producers with different problems from those they find on returning to their home countries. One suggestion that has been made to improve this situation is that secondary-school leavers and/or recent graduates be obliged to spend some time in actual production units, in order to familiarise themselves with their problems. Such a recommendation was, for instance, made by the comprehensive employment strategy mission to Kenya.[3]

EXTENSION SERVICES: LINKAGES WITH PRODUCERS

The case of agriculture was quoted in the previous chapter as representing a particularly clear case of the need for adaptation. Similarly, in the case of

---

[1] UNCTAD: *Transfer of technology,* op. cit., p. 8.

[2] Bell et al.: *Industrial technology and employment opportunity . . .,* op. cit.

[3] ILO: *Employment, incomes and equality . . .,* op. cit. The report also suggested curriculum changes in the training of engineers.

new agricultural technology, the need for an extension service or similar mechanism to reach the actual farmers with the new technologies is particularly clear. Many or most developing countries have an agricultural extension service, even though it may lack resources and effectiveness and often concentrates on the larger farmers to the neglect of the poorer farmers. What is not yet so fully realised is that there is an equally urgent need for an industrial extension service to reach the smaller and informal sector firms, not only to assist in adaptation but also to familiarise them with new technology. Really effective industrial extension services exist only in relatively few developing countries.

Several research institutions have set up teams of industrial liaison workers for extension work. Experiences with these and other methods of creating links between research institutions and the actual producers would be well worth studying. Similarly, there is an impression and presumption that a decentralised system of smaller-scale, more locally or regionally[1] oriented research institutions would find it easier to be effective in translating R and D into improved productive technologies. This would also be worth exploring, although one must leave room also for the possibility that smaller scale and decentralisation may hamper the work and make the pursuit of consistent national technology plans more difficult.

The R and D institutions should give special priority to providing assistance in building up effective capital-goods industries and also to developing engineering and design capacities throughout the country. The reasons for these priorities were explained earlier in this chapter. All this does not mean that there will no longer be a need to strengthen also the organisation of R and D institutions; but any such strengthening should be based on a clear recognition of their functions and should be considered as only one element in a technological system. Such a system would include, apart from central and decentralised R and D facilities, the development of engineering design facilities, extension services and a communication system between producers and the R and D sector, the encouragement of innovation at the factory level, a special emphasis upon sectors and industries that are essential in a basic-needs strategy, the appropriate training of the various types and levels of manpower needed, both at the production and at the R and D levels. All these elements must be welded together into a coherent technology plan, which in turn should be an integral part of the development plan. Even this brief list indicates that this will be a formidable task. Yet the technology gap at present is so great and the cost of technological dependence is so high that no time should be lost in changing the present uneven distribution of technological capacity.

The relationship between a central R and D institute and in-plant innovation and adaptation is similar to the wider relationship between central planning and community participation. In technology policy, just as in

---

[1] In the sense of regions within a country.

national planning, increasing recognition is being accorded to the importance of a two-way feedback relationship between the planners and the "planned", and to attempts to devise participating institutions and other links. This similarity underlines once more the integral relationship between national planning and technology planning. Governments should promote and subsidise appropriate in-plant R and D of a formal or informal nature as much as they should establish a national R and D structure. But the social value of in-plant R and D and innovation generally depends on a coincidence of private and social profit, and this in turn depends on the price and other incentives provided by the government. Similarly, the public interest may also require that the innovation be disseminated from the company where it was made to other firms or other sectors of the economy (or even to other countries, if an agreement to share technology exists). This again requires confidence and a sense of participation by the firms involved, and may involve special incentives.

Among the participating institutions designed to provide the desired two-way feedback is the research association, financed by the industry itself but with matching funds provided by the government. Another variation is the R and D institution financed by a levy on the firms in the branch served. Apart from stretching government funds, the hope is also that the fact of having to pay at least a share of the cost of the institution will make industry readier to bring their problems to the institution which they consider as "theirs".

Another essential intermediary, the engineering design sector, has been mentioned already. Specialised consulting firms in this field can provide design and related services with especial concentration and economy. As mentioned before, over a certain area where the appropriate technology is within the range of existing knowledge, engineering design skills are needed, rather than R and D. In the context of a basic-needs strategy, product design may become particularly urgent, because the existing product may be inappropriate although the need covered is essential. Another priority area is "descaling" design for small-scale, rural and informal sector producers. The use of local materials will often also require both product and process design skills. The apex of the engineering design machinery will be national design centres.

The Report of the Director-General of the ILO to the World Employment Conference adds four further formulae arising from past experience as elements favourable to the success of R and D institutions:[1]

(a) a fairly high degree of specialisation by product group or sector (for example, the International Rice Research Institute); technological know-how is rarely interchangeable between products except at a very general level of engineering or scientific principles;

---

[1] ILO: *Employment, growth and basic needs...*, op. cit., p. 149.

(b) close links with similar institutions in other countries which over the years have acquired a profound knowledge of the materials and processes involved if not of the particular socio-economic setting in which they need to be combined by the developing country institution; the UNDP has proposed sponsorship of such "twinning" arrangements to encourage an exchange of know-how and personnel;

(c) close relationships with the users of the technology developed; this may best be stimulated by including representatives of user industries on the governing boards and by charging a levy on member firms of employer associations to ensure their financial interest in the institution's work, and

(d) an appropriate salary and incentive structure for the staff which rewards the development of practical applications rather than theoretical investigations.

BASIC AND APPLIED RESEARCH

The presence of a communications gap between the R and D sector and production is accompanied by a tendency for R and D activities in the public sector to concentrate too much on pure research and to carry out insufficient applied research. In fact, it is stated that the share of pure research in the R and D sector is greater in the developing than in the industrialised countries.[1] This is certainly an undesirable inversion of what one should expect. However, it would be defeatist to draw from this fact the conclusion that developing countries should be advised not to try to build up their own R and D capacity. There are enough examples of successful R and D work in developing countries properly linked with and applied to actual production. Moreover, such advice would be in disregard of the claims of a basic-needs strategy, which requires products and processes that are often essentially different from those available from the technological shelf and often also beyond the scope of adaptation.

However, the advice given to developing countries to be cautious in competing in research with the industrialised countries[2] is basically sound. Even then it is subject to the provisos that: *(a)* basic research may be necessary and helpful as a part of the system of higher education and training for scientific personnel; *(b)* there are gaps in world basic research, especially in relation to the natural resources of developing countries, which it would be

---

[1] G. Ranis: "Some observations on the economic framework for optimum LDC utilization of technology", in L. J. White (ed.): *Technology, employment and development,* Selected papers presented at two conferences sponsored by the Council for Asian Manpower Studies (1974), p. 89.

[2] R. Nelson: *Less developed countries, technology transfer and adaptation, and the role of the national science community,* Yale Economic Growth Center, Discussion Paper No. 104 (New Haven, Conn., 1971).

strongly in the interests of developing countries to fill, in order to strengthen them in their competition with synthetic substitutes; and *(c)* there is some evidence that applied research may be less effective (for example, in agriculture) in the absence of some local basic research capacity.[1]

IMPACT OF HIGHER OIL PRICES

The impact of higher oil prices illustrates the acute need for technological capacity. Naturally, it raises also the question of new criteria for the selection of technologies from the shelf as well as questions of adaptation, but the problem transcends such relatively easier adjustments. The development of energy-saving products and processes emerges as vitally important, and on the whole it fortunately coincides with, and strengthens, the requirements of a basic-needs strategy. The search for alternative sources of energy, especially those useful for small-scale units of production and rural and scattered markets (such as solar power or bio-gas units), will be greatly encouraged, and it is important for developing countries to play their part in the application and development part of this research. Inevitably, this also involves some share in the basic research, at the very least in the sense of familiarity with such basic research. If it is argued that one can rely on the industrialised countries which are also affected by the rise in oil prices to produce the energy-saving new technologies, the danger is that the industrialised countries will find their own appropriate solution. This may well mean that the incorporation of energy-saving features will make future equipment and manufactured goods (as well as synthetic substitutes) even more expensive, and hence technologies on the shelf will be even more capital-intensive. (The same is also true as regards pollution control and other environmental requirements.)

The rise in oil prices may also give some of the natural products of developing countries a new advantage over synthetic substitutes. It will amply pay to utilise and develop these natural products with the help of new and improved technologies. On the other hand, the new balance-of-payments pressures to which the non-oil-producing developing countries are now exposed will make it necessary for them to reduce payments for the transfer of technology, to substitute national production for a number of previously imported items and to promote exports, thus making new claims on indigenous technological capacity. The substitution of home production by foreign investors for imports will serve to solve the balance-of-payments problems created by higher oil prices only if conditions are reasonable and if the profits transferred out of the country are not excessive.

Some World Employment Programme research has already been

---

[1] See R. Evenson: "Technology generation in agriculture", in Lloyd G. Reynolds (ed.): *Agriculture in development theory* (New Haven, Conn., and London, Yale University Press, 1975).

reoriented towards some of these important new problems.[1] The desire to reduce oil costs could be an element in favour of substituting labour for machinery. This was in fact shown to be the case in another World Employment Programme study of road construction in Thailand.[2] Yet another study, relating to the manufacture of cement blocks in Kenya,[3] also shows that fuel costs per block of cement are much higher in the case of large stationary vibrating machines than in that of small machines.

PRIORITIES IN TECHNOLOGY POLICY

There is legitimate cause for differing opinions as to whether, in the years to come, the main thrust in building up the technological capacities of developing countries should be within the production units, with an emphasis on training, management and motivation involved, or on the building up of R and D institutions with proper links with the production units. Perhaps the question does not arise. Obviously both elements of technological capacity are vital and must be strengthened, both individually and, particularly, in their interaction with each other. This calls for an over-all view and a planned distribution of resources devoted to strengthening technological capacity. What is certain is that the strengthening of national institutions is a matter of high priority for the governments of developing countries and for such proposed new instruments of international collaboration as a Consultative Group on Appropriate Technology or an International Appropriate Technology Unit (both of which were discussed at the World Employment Conference[4]), or the regional and subregional centres supported by UNCTAD and the World Employment Conference. Clearly, one way of strengthening national units is by keeping them in touch with each other's progress and by disseminating the results of research not only on a national but also on an international basis. But there can be no doubt that the linkages between the R and D level and the production level are perhaps the most vital links in the chain—including the development of design capacity, a capacity to use extension services effectively, the development of efficient consultancy services and, perhaps most of all, the development of national capital-goods industries.

---

[1] See ILO: *World Employment Programme: Research in retrospect and prospect,* op. cit., pp. 22 and 96-97, with reference to the study on the technological and employment implications of higher oil prices in Sri Lanka.

[2] ibid., p. 117, and McCleary et al.: *Equipment versus employment...,* op. cit. The effect of fuel prices combined with utilisation rates of equipment is shown to be decisive when comparing the social value of capital-intensive and labour-intensive methods.

[3] See F. Stewart: "Manufacture of cement blocks in Kenya", in ILO: *Technology and employment in industry,* op. cit., pp. 225 and 312.

[4] ILO: *Employment, growth and basic needs...,* op. cit., pp. 150-154, for details of the proposed Consultative Group and Unit. At the World Employment Conference this proposal was supported by the Group of 77 (the developing countries) and the Workers' group but failed to win the support of most of the industrialised market economy countries.

The development of design capacity can be combined with the development of efficient consulting services by fostering consulting organisations. These may be in the public or in the private sector. Special attention should also be given here to the problems of small-scale rural and informal sector producers who will often require more general, less specialised services. Foreign consultants will not usually have the skills or motivation (which national consultants could acquire) for designing descaled and more labour-intensive processes. For this purpose national capacity will be needed. The same applies to adaptation. The different management problems of new labour-intensive technologies create a need for effective management consultancy, where once again national consultants could acquire a greater understanding than foreign consulting firms of the problems associated with appropriate technologies specific to the developing country. Experienced consultants would also form a part of the communications network linking the R and D institutions with the production units. Consultants should be informed of R and D progress and possibilities to bring to the attention of their customers, and they in turn would be in a good position to bring to the attention of the R and D institutions those production problems which seem to require new R and D work, and also the opportunities they see in their work to increase producers' knowledge of the results or potentialities of R and D. Consultancy services have been described as "a vital part of any development strategy which seeks to improve the technological base of the production process".[1]

INTERNATIONAL ACTION IN THE CREATION OF NEW TECHNOLOGY

The emphasis in this chapter has been on the development and strengthening of national capacity for technological innovation. Nevertheless, there is obvious scope here for international support and co-operation. A number of proposals have been made, from the general proposals in the United Nations World Plan of Action for the Application of Science and Technology to Development, the International Development Strategy for the Second United Nations Development Decade and the more recent proposals by the Sixth and Seventh Special Sessions of the United Nations General Assembly, to specific proposals presented at the Fourth Session of UNCTAD and the World Employment Conference. There is broad agreement about the objectives and need for intensified international co-operation, but less so with regard to the precise instruments to be developed.

The resolution on strengthening the technological capacity of developing countries,[2] adopted with general approval at the Fourth Session of UNCTAD

---

[1] V. V. Bhatt: "On technology policy and its institutional frame", in *World Development,* Sep. 1975, p. 657.

[2] Resolution 87(IV) of 30 May 1976.

(Nairobi, May 1976), recommended inter alia specific measures of co-operation among developing countries at the subregional, regional and international level, including preferential agreements for the development and transfer of technology among themselves; the establishment of subregional and regional centres for the development and transfer of technology; the exchange of information; common technological research and training programmes; the provision of assistance to national centres or authorities on such matters as information on sources of technology and the preparation of model contracts for licensing agreements, and so on. The co-operation of the developed countries is requested on a number of specific matters, including the control of restrictive practices limiting the transfer of technology; encouragement of their enterprises and institutions to develop and to disseminate, on equitable terms, technology appropriate to the needs of developing countries; assistance in training by special programmes in both developed and developing countries; strengthening of global research capacities for dealing with specific and critical sectors of particular interest to developing countries; and expansion of their own R and D activities to promote technologies that are suitable to the requirements of developing countries.

The Programme of Action adopted at the World Employment Conference also contains relevant recommendations. The section on "Technologies for productive employment creation in developing countries" supports the establishment of national, subregional and regional centres to promote co-operation both between developing countries and between the latter and developed countries, and asks the ILO to help in the establishment of these centres (para. 52). It also calls on foreign firms to assist in the development of national technological capacity and in introducing and adapting "technologies which are both growth- and employment-generating, directly and indirectly"; and to "contribute to financing the training of national managers and technicians for the better utilisation and generation of technology" (para. 55). The Programme of Action asks the ILO to strengthen and reorient its own programmes in the collection and dissemination of information on appropriate technologies (especially for the rural sector), in manpower training and human resources development, and in research and technical co-operation in the field of development and transfer of technology (paras. 59-61).

The developing countries are concerned to ensure that international co-operation should not indirectly serve to continue or strengthen technological dependence; rather, it should be directed towards building up and supporting national capacities. This is in any case required by the country-specific (often indeed local-specific) nature of appropriate technologies, and the need (already emphasised) for close contacts with the intended users of the improved technologies. Moreover, shifting as much of the work as possible to national institutions builds up their organisational capacity and "experience by learning". The device of "twinning" has already been mentioned as a method of international co-operation which UNDP has proposed to sponsor. It has obvious potential, notwithstanding some obvious problems.

The United Nations World Plan of Action proposed two specific international targets for the creation of new technologies: *(a)* the industrialised countries should reorient their own R and D activities to the extent of 5 per cent of non-military R and D towards the problems of developing countries (at 1970 prices this would have required resources valued at US$2,250 million per annum; at present prices, the figure would be of the order of US$4,000 million per annum); *(b)* another sum of about half this amount (the target being 0.05 per cent of their gross national product) should be devoted by the industrialised countries to supporting increased R and D capacity in developing countries by way of equipment, financial aid, technical assistance and experts[1]—in each case either bilaterally or through international organisations. It will be seen that the total resources required at current prices would be of the order of US$6,000 million per annum. Such figures (which include private as well as public resources) must be seen in relation to the minimum (direct) cost of transferred technology at present paid out by developing countries. This was earlier given as US$1,500 million in 1968, quoting UNCTAD figures.[2] At present prices this would cost about US$3,000 million. If the assumed growth of such expenditure at 20 per cent per annum, as well as the indirect cost, through transfer pricing and so on, are included, the total cost of paying for transferred technology may well be in excess of the sums proposed as necessary transfers for the development of new technologies appropriate for developing countries.

Even if these precise orders of magnitude are modified—and the quantitative targets on which they are based have not so far been internationally accepted—this obviously raises the question of a mechanism for mobilising the resources required and determining priorities and channels. At present, no such mechanism exists, except in the agricultural sector through the Consultative Group on International Agricultural Research (CGIAR).

Both the over-all amount needed and the 2:1 distribution (between the reorientation of industrial country R and D and assistance to developing country R and D) of the total sum required can be questioned: indeed, so can the very principle of establishing targets or determining an *a priori* distribution of resources between the two main lines of action. However, the need for a new initiative and a continuing framework for encouraging the development and dissemination of appropriate technologies in the secondary and tertiary fields was clearly expressed in the UNDP statement to the World Employment Conference, as well as the need to provide a proper habitat

---

[1] The United Nations World Plan of Action suggested a ratio of 60 per cent for financial aid and equipment and 40 per cent for experts and related technical services.

[2] See Chapter 3.

within the United Nations system for efforts to promote appropriate technologies.[1]

The United Nations World Plan of Action also suggested a number of priority areas for new research and for the application of existing knowledge, but these, together with the rest of the Plan, are at present being reviewed in preparation for the United Nations World Conference on Science and Technology in 1979. The priority areas were, of course, defined before the seriousness of the oil and energy problem became apparent. The proposals for international action in support of new technologies would, of course, be additional and complementary to international action aimed at improving access to, and transfer of, existing technology. Proposals to this end have been put forward (and some initial action has already been taken) by UNCTAD, UNIDO and other United Nations organisations.

In the United Nations World Plan of Action it was proposed that a special World Plan of Action fund, or account, be set up within the framework of UNDP in order to act as the necessary catalyst and co-ordinating instrument. This fund would absorb 50 per cent of the additional contributions to UNDP projected between 1970 and 1975 and would amount to 25 per cent of the total resources of UNDP (in addition to any current UNDP resources allocated to relevant projects under the country programming and global projects provisions). However, no such fund or account has been established and UNDP has since developed on somewhat different lines. The question of the location and financing of the necessary catalytic and co-ordinating function has therefore still to be decided.

International action for the dissemination of information and the promotion and development of appropriate technology within developing countries and outside could go a long way towards assisting the developing countries to achieve greater technological self-reliance. The experience gained in agriculture has demonstrated that international efforts in promoting R and D and the co-ordination of efforts can be successful. The Consultative Group on International Agricultural Research, founded in 1971, has mobilised funds from international funding agencies, governments (bilateral donors) and non-governmental sources, and has promoted agricultural research through the establishment of international agricultural research centres and through financing R and D in existing national and regional institutions. However, no similar mechanism yet exists for financing R and D in the non-agricultural sectors, although it is felt by many that the need there is as great and that similar arrangements to cover the industrial and construction sectors could help to fill the present gap between developed and developing countries in the distribution of research on problems of industry. For these reasons, a proposal was placed before the World Employment

---

[1] Statement by Mr. I. G. Patel, Deputy Administrator of UNDP, who also indicated that UNDP was anxious to play a much more active role in the development and dissemination of appropriate technology (as proposed in the United Nations World Plan of Action).

Conference to create a Consultative Group on Appropriate Technology. The feeling that international action of some kind is needed is now widespread, and the Group of 77 developing countries and the Workers' group of the Conference endorsed the creation of such a body. The Workers' group emphasised that it should be tripartite in character, encompassing representatives not only of governments but also of employers and workers. However, most Western industrialised countries did not support this proposal to create a Consultative Group on Appropriate Technology.

Other proposals have come from various sources, and the resolution on development and international economic co-operation adopted at the Seventh Special Session of the United Nations General Assembly[1] suggested the creation of three international institutions, namely: *(a)* an industrial technological information bank, with possible regional and sectoral banks; *(b)* an international centre for the exchange of technological information for the sharing of research findings relevant to developing countries; and *(c)* an international energy institute to assist all developing countries in R and D for energy resources. Of these three proposals the first relates more to the flow of information on available technologies discussed in Chapter 3, and was mentioned there in its context. Moreover, the proposal is not specifically geared to the technology required by the small-scale or informal sector or under a basic-needs strategy; on the contrary, it is geared "in particular [to] advanced technologies". The third proposal relates to the single (although vital) priority area of energy (discussed above).

In addition to discussing the creation of a Consultative Group on Appropriate Technology, the World Employment Conference considered the question of the establishment of an International Appropriate Technology Unit. Such a unit, if created, would correspond at least in part to the second item mentioned in the resolution on development and international economic co-operation, which is cited above. The resolution speaks of "relevant" research findings, and the idea underlying the proposal to establish an International Appropriate Technology Unit was that it would be centred on promoting and sponsoring on a national basis the actual development of appropriate new technologies, in addition to sharing and diffusing the results of such research. However, it would be inappropriate to speculate whether the second proposal of the United Nations General Assembly would be satisfied by the creation of such an International Appropriate Technology Unit.

At the World Employment Conference the Group of 77 endorsed the establishment of such a unit, as did the Workers' group, which again emphasised that such a unit should be tripartite in nature. Most Western industrialised countries did not support the proposal for the establishment of an International Appropriate Technology Unit, just as they had not supported the proposal for the creation of a Consultative Group.

---

[1] Resolution 3362 (S-VII) of 16 Sep. 1975.

Other related proposals have been put forward, such as that of the United States National Academy of Sciences, included in the United States proposals to the Seventh Special Session of the General Assembly and to the Fourth Session of UNCTAD, for an International Industrialisation Institute to sponsor and conduct research on industrial technology together with the governments, industries and research facilities of developing countries. UNDP has recently proposed the establishment of international R and D institutes for specific industries or small groups of closely related industries.[1] Other proposals in the same vein could be mentioned. Certainly the multiplicity of similar proposals coming forward at the present time indicates the existence of a widely felt need and of a gap to be filled.

Much of the necessary international co-operation could be among the developing countries themselves. This would reduce some of the uncertainties and heavy costs that inevitably arise when an individual developing country (except perhaps the largest and technologically most developed) goes ahead on its own and risks spreading its own limited R and D resources too wide and too thinly. Instead, each country could specialise in building up "centres of excellence" in fields suggested by the special interests of a given country, or by its special potential for research. Such a deliberate blending of creation and transfer of new knowledge would seem to correspond closely to the aim of "collective self-reliance".

International support for the development and dissemination of appropriate technology need not entirely depend on government action. A number of non-profit-making organisations such as the Rockefeller Foundation and the Intermediate Technology Development Group in London are active and have a considerable impact in this field. So are a number of large companies, especially those with major multinational connections with developing countries, of which Philips with its special pilot plant in Utrecht is perhaps the best-known example. Beyond this, the contributions of multi-national enterprises to the R and D work and infrastructure of their host countries are a subject for considerable current debate and proposed action. By transferring research to their subsidiaries in developing countries, multi-national enterprises could transfer to developing countries some of the effects of R and D on learning and training, provide resources for research on more appropriate technologies and relieve the balance-of-payments problem by reducing transfer payments for technical services to the parent company and by retaining the cost of R and D expenditures within the country. Apart from carrying out some of their own R and D in the developing country, multinational enterprises could also use some of their tremendous R and D capacity to help their suppliers and distributors in the developing countries with R and D, either at their head offices or preferably in the developing

---

[1] UNDP: *The future role of UNDP in world development in the context of the preparations for the Seventh Special Session of the General Assembly: Report of the Administrator* (New York, doc. DP/114, 24 Mar. 1975; mimeographed), paras. 36-37.

countries, perhaps in the suppliers' and distributors' own plants. The risk is that R and D carried out by multinational enterprises may not be readily available to national producers, although this point could also be covered in agreements with host governments and in the proposed code of conduct or guidelines for multinational enterprises.

Another risk is that the R and D located and disseminated by multi-national enterprises in the developing countries might take the form of research on inappropriate products or inappropriate (for example, capital-intensive) processes, leading to an internal brain drain (that is, one diverting scientific resources away from national needs) and possibly serving as a step towards an external brain drain (that is, transferring to the head office any promising and well trained scientists from the local R and D establishments who may have caught the eye of the head management). In other words, the transfer of R and D work to the subsidiaries of multinational enterprises in developing countries is perhaps most useful if it takes place within the context of vigorous national R and D development in the host country, geared to the country's own problems and providing a satisfactory working environment for trained local staff.

It has also been pointed out that the changes in demand patterns and relative prices resulting from a basic-needs strategy, especially when combined with social cost pricing and the removal of factor price distortions, would go some way to ensuring that the R and D work of multinational enterprises would be redirected to more appropriate products and processes.

In the proposed code of conduct for direct foreign investment, and in the recommendations, guidelines and training efforts designed to secure more effective bargaining and negotiations between developing countries and multinational enterprises, provisions for a greater contribution by multi-national enterprises to the technological capacity of developing countries, including R and D capacity, play a considerable part. The best method of ensuring such a contribution and the form it should take will differ from country to country, from industry to industry and even from one multinational enterprise to another. But certainly there should be room to increase the present share (almost certainly well under 1 per cent) of the R and D expenditures of the multinational enterprises in the developing countries in which they operate. An extension of local subcontracting and the use of local resources would help to give additional incentives in this direction. But the extension of local subcontracting and the development of networks of local suppliers, in their turn, depend to an important degree upon the technological capacity of such local units. Thus, once again, the relationship between the development of national capacity and the ability of securing from international contacts good terms leading to a further transplantation of technological capacity are seen to be complementary and cyclically connected. In the light of the previous discussion on the importance of a national capital-goods industry for national technological capacity, the building up and use by multinational enterprises of national engineering (machine building, repair and maintenance) services would be especially important. The same applies to the role of multinational engineering enterprises.

The main problem will be to direct the potentially large contribution of the multinational enterprises to appropriate R and D towards the small-scale and informal sectors. This will not come naturally to them, since their natural interest is in larger-scale, capital-intensive processes and standardised, often high-income goods. It is partly for this reason that the possibility of increasing the tax contributions of multinational enterprises to their host countries has been mooted, with a view to providing increased financial support by multinational enterprises to the national R and D structures of the countries in which they operate.[1]

---

[1] See A. S. Bhalla and F. Stewart: "International action for appropriate technology", in ILO: *Tripartite World Conference on Employment . . .*, op. cit., Vol. II: *International strategies for employment.*

SOME PROBLEMS OF TRAINING AND EDUCATION FOR APPROPRIATE TECHNOLOGY

6

TRAINING

The question of whether capital-intensive methods displace skills, or on the contrary are complementary with skill intensity, has been hotly debated. It is obviously an important question. Developing countries suffer from shortages of certain kinds of skill as severely as, or more severely than, from a shortage of capital. Labour-intensive methods which place a heavy strain on very scarce skills may therefore in fact be inappropriate—except to the degree that one may assume that widespread employment will lead to "learning by doing", and thus in effect help to form and create the necessary skills.

Research into this question has been going on for some time, although not as intensively as the importance of the problem would warrant. The results have been inconclusive in the sense that no general answer has emerged: it is not possible to state in a general sense that labour intensity saves skills or that it requires more skills than capital-intensive technologies. Very probably, the real answer is that capital- and labour-intensive technologies require different *kinds* of skill, and it is difficult to weigh the different kinds of skill to arrive at judgements in terms of "more" or "less". Moreover, research has shown the danger of generalisation: different concrete cases yield different results, depending on the nature of the industry, the scale of operations and the type and level of skills concerned.

A particular kind of skill required for larger-scale labour-intensive operation is skill in supervision. This applies not only in large-scale industrial and agricultural operations but also in construction and labour-intensive public works. Multinational enterprises may have more access to supervisory skills than comparable national producers. For example, the comprehensive employment strategy mission to Kenya found that, wherever foreign and local manufacturing companies could be meaningfully compared, contrary to expectations the foreign companies tended to use more labour-intensive methods; this was attributed by the mission to superior supervisory skills.[1]

---

[1] ILO: *Employment, incomes and equality...*, op. cit., technical paper 16.

Where such skills are absent, producers have to rely more on machine-pacing of labour performance and this leads to capital intensity. However, other studies (as well as the report of the Kenya mission) have also found that multi-national enterprises tend to pay higher wages, which would give an incentive to more capital-intensive methods; and they certainly have easier access to capital. Moreover, while the large-scale multinational enterprises may have superior supervisory skills, the very small-scale and informal sector operations may have no problem of supervision at all, particularly in cases of self-employment and family modes of production. Capital-intensive methods require high-level technicians to maintain the machine processes but perhaps fewer operative skills and fewer foremen and supervisors. But, on the other hand again, they require a more disciplined and punctual labour force, and any breakdown of machinery due to carelessness or ignorance can be very costly.

As will be seen, the mixture and combinations of the different kinds of skill required do not lend themselves to easy generalisations. A World Employment Programme study in Kenya has shown that automated (capital-intensive) processes require more supervisory skill but less of other kinds of skilled labour in the actual operation, so that there is a choice of combination between highly skilled plus unskilled labour in the automated processes and semi-skilled labour in the semi-automatic (less capital-intensive) processes. It is obviously not possible to say whether a transition to less capital-intensive processes "saves skills" or not. The very question appears oversimplified. Other studies suggest that skill requirements may be relatively high with both capital-intensive and labour-intensive technologies, but relatively low with technologies intermediate in degree of capital intensity. World Employment Programme research in Thailand has shown that labour-intensive and capital-intensive processes also involve different kinds of supervisory skill. Probably the skills needed for organising workers are more readily available in developing countries than those needed for automated lines.[1]

In industry (as distinct from agriculture) it has been found that it is in the supervisory and managerial skills of internal co-ordination and control that the formation of human competency has often proved a bottleneck, preventing the development of low-cost competitive industries, rapid technological improvement and full utilisation of capacity.[2]

The training of managers and engineers is also an important factor in the choice of technologies. It has been suggested that engineers, by virtue of their training and outlook, tend to select more capital-intensive techniques than commercial managers. Moreover, there is some evidence that managers with technical training and a background in production may be more likely to have a knowledge of alternative techniques and an understanding of the

---

[1] Cooper et al.: "Choice of techniques for can making...", op. cit., p. 112.

[2] See P. Kilby: "Hunting the heffalump", in P. Kilby (ed.): *Entrepreneurship and economic development* (New York, The Free Press; London, Collier-Macmillan, 1971).

values of labour-intensive processes than managers trained in sales and finance who depend on machine salesmen and external consultants. These findings are not necessarily irreconcilable, but their existence once again shows the difficulty of easy generalisation. In many studies, however, the skill pattern of the labour force and of the management has emerged as an important determinant of technology.

At its International Centre for Advanced Technical and Vocational Training in Turin and within its training programme, the ILO organises seminars on the training of managers and engineers in the choice of alternative technologies. One such seminar (on employment promotion training for managers, engineers and technologists in manufacturing and construction) was organised in July 1975 under the joint auspices of the Turin Centre and the World Employment Programme. It is also proposed to draw up guidelines for training managers in the choice of appropriate technologies. In general, managers are trained in choosing between competing variations of the same technology. To enable them to make a wise selection from a range of technologies, suitable evaluation methodologies and training in their judicious use need to be disseminated to managers and government officials with responsibilities in economic planning, environmental and resource conservation, the licensing of new industries, and so forth. All this is envisaged as a response to the recommendation of the United Nations Advisory Committee on the Application of Science and Technology to Development (ACAST) that "national and international organisations should make special provision . . . for educating engineers and managers in the appropriate choice of technology in developing countries and its adaptation to local conditions". [1]

In the field of management and supervisory training, programmes and course materials and even training methods are already in existence. An excellent illustration of this type of training for small-scale rural entrepreneurs is the Small Business Management Programme developed by the Province of Saskatchewan in Canada. Other programmes of varying sophistication are available for all levels up to the top echelons of large-scale industries. The problems are, however, those of the preparation of suitable trainers and the development of channels for delivering this training to those who need it. A substantial amount of experience in this area has already been acquired in the ILO as a result of technical assistance projects, but there are still some gaps, notably as regards rural training and training for the urban informal sector. These are the areas on which future action should concentrate (as discussed below).

Another field in which more work needs to be done is with regard to the appropriate management techniques for mobilising large numbers of workers employed on a temporary basis, a case which arises with large-scale labour-

---

[1] UNIDO: *Statement by the United Nations Advisory Committee on the Application of Science and Technology to Development at its Twentieth Session held at Geneva, 21 October- 1 November 1974,* Paper submitted to the Second General Conference of UNIDO, Lima, 12-26 Mar. 1975 (doc. ID/Conf./3/11, 19 Nov. 1974; mimeographed), p. 5.

intensive construction projects. While some experience has been gained in this area, [1] it would be useful systematically to classify and analyse management techniques for large labour forces for widespread dissemination for the purposes of training as well as implementation.

As has been mentioned, appropriate technologies have the advantage that they often provide on-the-job training and facilities for learning by doing, with beneficial effects not only for the long-term efficiency of the operation itself but also indirectly for the rest of the economy; a higher degree of on-the-job innovation may even result. By contrast, subsequent productivity gains with capital-intensive technologies are more likely to lead to the acquisition of more equipment and thus to use up scarce investment resources and foreign exchange.

The experience of African countries has shown the great significance of training by expatriate firms and how this training is transmitted to the rest of the economy through indigenous systems of masters and apprentices. It has also been observed that firms run by expatriate minority residents (e.g. Lebanese and Greeks in some African and Latin American countries, Indians and Chinese in Asian and other countries, and so on) play an especially important role in on-the-job training, since they are typically disposed to use appropriate and labour-intensive technologies.

World Employment Programme research into metalworking in Ghana has shown that master journeymen (heads of informal sector firms) who had worked in the modern sector as apprentices were more innovative than the ones whose training occurred only within the informal sector.[2] Other similar studies have also highlighted the importance of skill flows between the modern and informal sectors in the diffusion of technology.

EDUCATION

Behind such specific problems of the relationship between skills and appropriate technology lies the more general need to relate the whole system of education at all levels to the concrete problems of the country. An important feature of any policies to make the educational curriculum relevant to planning objectives and to a basic-needs strategy is the introduction of polytechnic elements into education. The World Employment Programme research project on education and employment is currently investigating a number of related questions: curricula and teaching methods appropriate to a basic-needs strategy; the educational systems most suited to the needs

---

[1] See, for example, M. Allal and G. A. Edmonds: *Manual on the planning of labour-intensive road construction* (Geneva, ILO, 1977), Ch. 10: "Organisation and management"; and E. Costa et al.: *Guidelines for the organisation of special labour-intensive works programmes* (Geneva, ILO, 1977; mimeographed).

[2] WEP research project by A. N. Hakam on technology diffusion between the modern sector and the informal sector, with particular reference to the case of metalworking in Ghana.

of the informal sector and rural workers; and the means by which teacher training can be upgraded to provide an effective implementation of the curriculum. A study in Bangladesh is designed to explore, among other factors, the influence of formal and non-formal education in the performance of farmers in the adoption and diffusion of technological innovations.

The unequal access to capital involved in inappropriate technology is mirrored by unequal access to education.[1] All too often, as the ILO comprehensive employment strategy missions and other studies have shown, the educational system alienates students from the real-life problems of the economy, in that it is based on imported curricula and methods and is geared to examinations that admit a student to the following stage of education, rather than catering for the needs of the great majority of students, for whom each level will be terminal. There is a high level of drop-outs, and the education of women and of those living in rural and remote areas is still inferior. The importance for the development of the rural sector of improved facilities for educational and vocational training is stressed in the Programme of Action.[2] And at the highest (university) level valuable national resources, both human and financial, are being spent on producing graduates more attuned to the needs and traditions of far-away industrialised countries than to those at home.

The ILO comprehensive employment strategy missions have spelt out the changes required by the demands of a basic-needs strategy. Science education, teacher training and the design of new methods and curricula are the priorities stressed in the redesigned framework of primary, and especially secondary, education as comprehensive and terminal units. Technicians rather than technologists are required in large numbers. Some of these efforts are beginning to bear fruit, and many countries are now trying to feature basic science and vocational training more prominently in primary- and secondary-school curricula, and to develop basic-needs-oriented systems of vocational training and informal education.

The role of education has been further enhanced by the growing evidence and realisation that formal technical training plays a smaller part than was previously assumed and that experience and on-the-job training are the main vehicles for implanting new skills.[3]

At present, the ILO is making efforts, in collaboration with UNESCO and UNICEF, to introduce technological elements into basic education in West Africa. These are linked with other efforts to introduce new technology into the villages and to provide, at the same time, appropriate training for the maintenance of new pumps, outboard motors and other mechanical means for

---

[1] Particularly in rural areas, as emphasised in the Programme of Action (para. 20).

[2] Para. 10.

[3] See Peter Kilby: "Farm and factory: A comparison of the skill requirements for the transfer of technology", in *The Journal of Development Studies* (London, Frank Cass), Oct. 1972, p. 67; and J. Maton: "Experience on the job and formal training as alternative means of skill acquisition: An empirical study", in *International Labour Review*, Sep. 1969.

improving productivity in rural areas. This is being done in response to the general observation that the simple introduction of new techniques and materials into a rural area without provision being made for the necessary skills for maintaining the new equipment have proved wasteful in many instances in the past.

TRAINING OF TECHNICIANS AND INSTRUCTORS

The need for engineering and design skills, emphasised in earlier chapters, offers ample scope for training programmes. No technology plan, no network of institutions, no policy will work without qualified personnel.

The Programme of Action adopted by the World Employment Conference singles out the training of technicians for appropriate technology selection (para. 50). Here also, the industrialised country, with its extremely high degree of specialisation and long periods of training, does not necessarily offer the appropriate model or blueprint. For instance, in the capital-goods industries of developing countries the need will often be for multipurpose machinery and hence for non-specialised casting workshops. This will require different qualities and a different training. Moreover, the proportion of medium-level technicians and other personnel of similar level to highly trained personnel should be greater within a basic-needs strategy (for instance, there should be many more trained health workers and nurses than doctors, among doctors more general practitioners than specialists, and so on).[1]

Efforts need to be made to develop training programmes which provide rapidity of training, and especially retraining. The old style of apprenticeship, which is so prolonged that in effect it "locks" the worker into one trade (and, usually, one technology) for his active life, must be avoided. The "modules of employable skills" concept now being developed by the ILO is particularly interesting from this point of view. In co-operation with a wide range of national projects of vocational training, modularised training programmes are being developed for basic professions, taking into account the needs not only of large-scale industry but also of the informal and rural sectors.

In its technical assistance programme the ILO is also heavily involved with institutions training machinists, fitters, welders, electricians, mechanics, woodworkers, bricklayers, pattern makers and foundry operatives. It is important that the training needs of the small-scale and informal sectors should not be neglected, and valuable programmes in this very field have already been devised in small-industry institutes, crafts training centres, co-operative workshops, and so on.[2] Among the methods recommended are

---

[1] It may be added that in the industrialised countries too second thoughts are at present being expressed about the present model of science education, and many reforms currently being suggested would move the system in a direction which would make it more similar to that now advocated for developing countries.

[2] See ILO: *Employment, growth and basic needs...*, op. cit., p. 147.

mobile demonstration workshops that may be taken into rural areas[1] and specialised appropriate technology institutions where draughtsmen and design engineers are trained to make blueprints and models suitable for small engineering workshops. There is plenty of scope for institutional innovation and new ideas in training for appropriate technology, as well as for exchanges of experience among various countries. For Colombia it has been suggested that recent engineering graduates should be employed for a year or two in plants where the equipment they design can actually be built and tested,[2] and the comprehensive employment strategy mission to Kenya made similar suggestions to make recent engineering graduates familiar with labour-intensive and small-scale production processes. Another suggestion which could be useful in this connection is the establishment at each university of a technology development centre, as is the case at Kumasi in Ghana.

Rural vocational training is often unjustifiably neglected in relation to industrial training. The ILO Eighth Asian Regional Conference, held in Colombo in 1975, after identifying the lack of technical, administrative and managerial expertise as a major bottleneck for the implementation of rural development plans, pointed out that some training of farmers and their families in off-farm skills can greatly help to cope with the widespread problem of seasonal unemployment. At the same time, seasonal unemployment offers an ideal opportunity for training by using otherwise idle time. Rural youth might be influenced by proper training to take up a rural occupation instead of joining the stream of migrants to the towns. The ILO itself has concentrated on the training of instructors, in view of the multiplier effect which can be expected from such a concentration. It is clear that any such plans for expanded and reoriented rural training require suitably trained and oriented technicians at various levels for staffing and direction.

At the rural community level two types of training will be required. There will be a need for modern versions of the village blacksmith, able to rig up simple machines such as the cassava-scraper developed by the Intermediate Technology Development Group, London, and capable of repairing equipment of non-village origin. The other need will be to develop entrepreneurial resources on a community scale. There must be a communal awareness of, and readiness to undertake, the functions of marketing (visualising and actualising potential markets and potential resources), accounting (avoiding profitless ventures) and financing (for example, community savings banks). This calls for some rather innovative programmes of rural adult education.

Past failures in rural appropriate technology have shown the need to ensure that any innovations are both socially and technically appropriate. Thus suitably trained anthropologists may participate in teams on appropriate technology. Engineers need to be trained to design equipment that is suitable

---

[1] The use of mobile training services for the benefit of rural areas was also recommended more generally by the ILO Eighth Asian Regional Conference.

[2] Instituto de Investigaciones Tecnológicas, Bogotá: "Capacity of the engineering industry in Colombia", in ILO: *Technology and employment in industry*, op. cit., pp. 251-252.

for rural use, tolerant in materials and dimensions and capable of being repaired by village mechanics. And the participation of people who *really* understand commerce in its various aspects, in order to help rural enterprises to find and obtain access to the markets they can supply, is particularly important but hitherto totally ignored.

An important obstacle to the improvement of training in appropriate technology is the lack of suitable training material. This applies to all levels (training courses, technical schools, universities) and to all types of material (textbooks, manuals, audio-visual material, and so on). Such training material is needed mainly for use in educational and training institutions in developing countries, but not exclusively so. The same need arises in institutions in industrialised countries which train people from developing countries or those who plan to work in developing countries. The preparation of such material should be mainly a task for the developing countries themselves, but again the industrialised countries and the international organisations have a role to play, and international assistance generally is by no means excluded. This is a priority area, partly because of the strong multiplier effect that would result from the removal of this bottleneck, which now so greatly inhibits the development and dissemination of appropriate technology; and partly because otherwise the widely prevalent notion of the inherent superiority of the technology of the industrialised countries (large-scale, capital-intensive and energy-intensive) is bound to persist. Only the demonstration of equally or more efficient alternatives will help to demolish this notion—and this in turn requires material that can be used in the training of those involved with technological decisions at various levels.

The Programme of Action adopted by the World Employment Conference calls for the formulation and implementation of national training plans (para. 56). In doing so, it distinguishes the following four levels:

(a) middle-level technicians and skilled workers to be employed in the production technologies associated with the goods and services required to satisfy basic needs;

(b) professionals, technicians, managers and skilled workers to replace expatriate staff who at present apply advanced technology;

(c) professionals and technicians needed to manage research and studies undertaken by national and/or regional technological research bodies; and

(d) technicians, professionals and skilled workers, who should be assured of a measure of social status and incentives to prevent a brain drain, in order to promote the utilisation of technologies designed to achieve material and social objectives.

INSTITUTIONAL, ADMINISTRATIVE AND PLANNING REQUIREMENTS FOR APPROPRIATE TECHNOLOGY

7

INSTITUTIONAL REQUIREMENTS

In any developing country, and especially in the larger and technologically more active ones, there is a multitude of government agencies concerned with the promotion, dissemination and use of technology. The ministries of industry, agriculture, transport, public works and other sectors of the economy all have to choose technology, both in the sense of methods to be employed in the public sector directly under their control and those to be recommended, transmitted, induced or prescribed to private sector operators. Nor must it be forgotten that such decisions on technology are not limited to the so-called "productive" sectors. There is a health technology too, with decisions to be taken on matters such as the choice between the construction of a few capital-intensive urban hospitals or of a network of rural health clinics. Even where the decision to build hospitals has been taken, hospital equipment can be ordered from small-scale or even informal sector producers or from large producers, or imported. The hospitals themselves can be built by labour-intensive or capital-intensive methods and involve small contractors in various ways and to a varying degree. Even the ministries of foreign trade and finance, by deciding which imports to encourage, which home industries to protect, which exports to promote, which foreign exchange restrictions to impose, and whether to restrict or to relax credit, make decisions that affect the scope and chances of appropriate technology. And this impact is usually important enough—especially in the context of a basic-needs strategy—to be considered and weighed. Thus, in a sense, the promotion of appropriate technology involves the co-ordination of the whole government machinery.

In a more specific sense a whole network of institutions in the field of science and technology is involved in the implementation of an appropriate technology policy. One blueprint of such a network of institutions is included in the United Nations World Plan of Action for the Application of Science and Technology to Development. [1] Smaller developing countries and those

---

[1] See United Nations, Department of Economic and Social Affairs: *World Plan of Action...*, op. cit., pp. 89-91.

at the beginning of their technological development will of course not have the full set of institutions shown in this or other blueprints, and may wish to develop some of the required institutions on a regional rather than a national basis. The different traditions and different problems of developing countries will lead to a different emphasis and to different areas of specialisation. Most larger countries, in fact, will wish to add their own specialised institutions for research on their major export commodities—oil, rubber, coffee, and so on, as the case may be. It would also be possible to list and classify the various elements of such an institutional network in different ways. All the same, the United Nations World Plan of Action blueprint could serve as the basis for a rapid survey of the network of relevant institutions, in a way that would help to bring out the problems of interaction and co-ordination between them. The Plan classifies the institutional requirements as follows:

(1) national policy-making bodies in science and technology:
    (a) central science policy-making body;
    (b) R and D promoting and co-ordinating bodies;
(2) higher education institutions in science and technology:
    (a) science faculties in universities;
    (b) "third level" polytechnic schools and schools of engineering;
    (c) "third level" schools of agriculture;
    (d) "third level" schools or university faculties of medicine;
(3) technician training institutions:
    (a) technological training institutions;
    (b) agricultural training institutions;
    (c) medical training institutions;
(4) research and experimental development institutions:
    (a) fundamental research institutes;
    (b) applied research and experimental development institutes; and
(5) scientific and technological public service:
    (a) natural resources and environment services;
    (b) information and documentation services;
    (c) museums and collections;
    (d) standards, norms and instrumentation; and
    (e) extension and innovation services.

Each of these different categories is then broken down into the various types of more specific institution (72 in all).

This network list is not specifically geared to any concept of appropriate technology but rather to the development of over-all scientific and technological capacity. However, it is clear that all such institutions and all such developments of capacity should be governed by the concept that whatever is done must be appropriate to the requirements and conditions of the country concerned. However, even in the much narrower sense of institutions that are specifically geared to the development of a particularly labour-intensive, local-resources-using, energy-saving or otherwise specially appropriate technology, it will be found that there are a number of special institutions of this kind, outside the general network previously described.[1] In India, for instance, there is an Appropriate Technology Cell in the Ministry of Industrial Development, a National Small Industries Corporation, Small Industry Service Institutes, a Small Industry Extension Training Institute, a National Research Development Corporation, the Gandhian Institute of Studies (with an Appropriate Technology Development Unit), the Central Leather Research Institute in Madras, the Planning Research and Action Institute in Lucknow, and so on. The set of small industry institutions controlled by the National Small Industries Corporation alone has assisted 100,000 firms, trained a similar number of people and supplied machinery on hire purchase to a value of some US$80 million. The National Research Development Corporation has almost 1,200 registered inventions and has established prototype production units and pilot plants. A scaled-down process for sugar production developed by the Planning Research and Action Institute has been applied to about 900 units. Similar networks, although on a smaller scale, exist in a number of other developing countries, especially in Asia. In Africa the East African Industrial Research Organisation (covering Kenya, Uganda and Tanzania), the Technology Consultancy Centre at Kumasi University of Science and Technology (Ghana) and the Federal Institute of Industrial Research (Nigeria) are examples of institutions with considerable achievements in the field of appropriate technology. Latin American examples would include the Instituto Nacional de Tecnologia Industrial in Argentina and the Instituto de Investigaciones Tecnológicas in Colombia. There are also important institutions in the industrialised countries, such as the Intermediate Technology Development Group in the United Kingdom, the Brace Research Institute in Canada, the Tool Foundation in the Netherlands and, in the United States, Volunteers in Technical Assistance as well as the Technology and Development Institute at the East-West Center in Hawaii.[2]

The OECD Development Centre has suggested[3] that institutions of this kind that are more specifically geared to appropriate technology can be divided into three big categories:

(a) higher education institutions, or rather, small groups which originated from and are closely linked with such institutions, such as Kumasi University of Science and Technology (Ghana), Mindanao State University (Philippines), the Universidad de los Andes (Colombia) or the Technische Hogeschool Eindhoven (Netherlands);

(b) governmental, private or semi-public organisations specialising in inter-

---

[1] See Appendix B, pp. 145-160.

[2] For a description of the activities of these organisations, see Appendix B.

[3] Nicolas Jéquier (ed.): *Appropriate technology: Problems and promises* (Paris, OECD Development Centre, 1976), p. 55.

mediate technology and working primarily on a national basis, such as the Appropriate Technology Cell in the Ministry of Industrial Development and the Gandhian Institute of Studies (India) or the Appropriate Technology Development Organisation (Pakistan); and

(c) multinational groups or research centres, such as the London-based Intermediate Technology Development Group (United Kingdom), Volunteers in Technical Assistance (United States) or the International Rice Research Institute (Philippines).

The classification presented in the United Nations World Plan of Action, mentioned above, distinguishes four functional levels: [1]

(a) planning, decision and control;

(b) co-ordination, promotion and financing of scientific and technological research at the national level;

(c) execution of research (operational network) through institutions for higher education in science and technology, technician training institutions and research and development institutions; and

(d) scientific and technological public services.

The 72 types of institution previously mentioned are then allocated to one or other of these four functional levels. Once again, it is possible to develop different criteria for establishing functional levels, or to allocate different institutions to other levels than those in the United Nations classification; but it is clear that, in one way or another, the four functions listed above must be carried out by the network of institutions and that each level must have its own and diversified institutional requirements and must be part of a coherent and consistent system.

The importance of the fourth functional level, the scientific and technological public services (STS)—which could also be described as the "infrastructure" of the institutional network—is too often overlooked in relation to the third level, the actual R and D level. The estimate in the United Nations World Plan of Action was that the institutions at the STS or infrastructure level would absorb the same amount of financial resources as the other levels combined. In view of the importance of extension and innovation services in the context of a basic-needs strategy, this may well be considered an underestimate.

The list of institutions given above does not include all those who make technological decisions of importance. The ordinary government departments and their technical sections have already been mentioned. Development financing institutions may be added—investment banking played an indispensable part in developing industrial technology in a number of European countries. This, of course, is quite apart from the units and decision-making agencies within the private production sector itself. There are often

---

[1] United Nations Department of Economic and Social Affairs: *World Plan of Action . . .,* op. cit., pp. 65-67.

definite advantages in linking technological advice or training with lending or with managerial training in book-keeping, stock control, and so forth, in courses or operations under the auspices of small business administrations.

The science and technology institutions that at present exist in most developing countries were created for the development and promotion of technology for the modern large-scale sectors. As noted above, the recent preoccupation with intermediate and appropriate technology especially suited to the needs of the small-scale informal producers has also led to the creation of some special institutions to deal with the subject. For example, in countries such as India and Pakistan, which seem to be the pioneers of the concept and movement of appropriate technology, the existing national institutions on scientific and technological research did not undertake the additional responsibility for the development of intermediate and appropriate technology. Instead, separate institutions were specially created and designed for this purpose, presumably on the assumption that it was difficult to reorient the existing scientific and technological institutions for upgrading rural and small-scale technology. Since the development of technologies appropriate to the small-scale sector is supplementary to the development of technology for the large-scale sector, it is essential that when the two types of technologies are promoted through parallel institutions, such institutions and systems should interact to the mutual benefit of both. It may be desirable to allocate the responsibility for the large-scale modern science and technology institutions with the central government while leaving the responsibility for the small-scale intermediate technology with local and provincial governments and private institutions.

So far, only national and local institutions have been considered. It would be possible to draw up a corresponding list of international and regional institutions and arrangements needed to support and co-ordinate the national institutions. If such a list were drawn up and compared with reality, many gaps would become apparent. A number of proposals to fill some of these gaps have been put forward in various contexts and from various sources. Two specific proposals of this kind of special importance were before the World Employment Conference.

In the preceding chapter it was pointed out that it is pointless to set up institutions unless trained and motivated personnel are available to make them work. It may now be added that the success of these institutions will also depend on the full co-operation of the scientific community, of advisory panels of leading citizens, of management and workers in the productive sectors and indeed of the community as a whole. Rewards for successful innovations and useful suggestions, publicity in newspapers, radio, television and so on may help here. This last point also leads back again to the importance of an educational system, both formal and informal, which inter alia creates an attitude of interest in technological problems and an understanding of its vital role in national development. Special and explicit attention to technology policy and the development of technological institutions would normally be indicated in any national development plan. The very exercise of doing this would bring home to planners the need for consistency in technology policy and development policy and would prevent implicit inconsistent decisions in these two fields to slip in unnoticed.

ADMINISTRATIVE REQUIREMENTS

Among the issues that arise when such a considerable network of institutions is administered, in any field, the contrasting merits of centralisation versus decentralisation feature prominently. Decentralisation has obvious advantages, to help to eliminate the communications gap discussed earlier in this book. This is particularly so in the context of a basic-needs strategy, when local needs and the nature of local poverty problems may differ greatly in different regions of a country, when the use of appropriate technology involves the participation of innumerable small production units, and in the context of rural development generally (where the obvious need for community involvement and grass-roots identification of problems has led to many variations of decentralised administration). Apart from ensuring closer links between real local or regional needs and R and D activities, regional and local institutions may do useful research in unfashionable yet important areas which are of regional but not of national importance and which would otherwise be neglected.

One of the chief co-ordinating tasks, and one making strong claims on the quality of administration, is the right "packaging" of technology with extension services, information and communication services, credit, the supply of inputs such as fertiliser or irrigation in the case of agriculture, price incentives, the structuring of demand by means of fiscal and other policies, and so forth. It is paradoxical that, whereas in the case of transfer of technology from abroad the big problem for the developing countries is that of "unpackaging" or "unwrapping the package", in the case of administering an appropriate technology it is precisely the opposite problem of proper "packaging".

The table[1] opposite illustrates both the political problems (cols. (2) and (3)) and the administrative requirements (col. (4)) for dealing with the basic-needs-oriented development strategy based on appropriate technology proposed by the ILO comprehensive employment strategy mission to Kenya. The administrative requirements of a basic-needs strategy are, of course, much wider than those of administering an effective technology policy *per se*; but, since appropriate technology will only be meaningful in the context of a broader basic-needs development strategy, they are not irrelevant. They illustrate the above-mentioned "packaging" problem.

---

[1] Taken from W. F. Ilchman and N. T. Uphoff: "Beyond the economics of labor-intensive development: Politics and administration", in *Public Policy* (Cambridge, Mass., Harvard University Press), Spring 1974. Reprinted in *Ekistics* (Athens, Athens Center of Ekistics), Aug. 1975, p. 93.

A political-administrative comparison of labour-intensive development policies

Policy (1)	Support from (2)	Opposition from (3)	Administrative requirements (4)
Agricultural extension for smaller farmers	Small farmers	Large farmers	Effective extension service willing and able to reach small farmers
Rural public works	Workers on projects; beneficiaries of works constructed	Persons who must pay higher taxes, wages, and/or prices	Organisation, design and management of widespread, dispersed construction projects
Industrial extension for small-scale and rural producers	Owners and managers of small-scale and rural enterprises	Owners and managers of competing large-scale enterprises	Extension service linked with "appropriate technology" R and D reaching small producers
Public housing	Workers employed; sellers of building materials; recipients of public housing	Owners of rental housing	Organisation of construction and financing of houses; management of public housing
Tariff revision of duties and quotas	Producers of import substitutes	Consumers of capital goods and luxury imports	Effective import controls to prevent smuggling and black market
Regional distribution of investment	Regional and rural sectors generally	Property owners and residents of metropolitan areas	Inducement of investors to invest in outlying regions; control over metropolitan investment
Income policy—freeze at top	Lower income groups	Upper income groups	Control over wages and salaries paid in formal sector
Progressive land tax	Intelligentsia (?)	Large landowners	Capability of assessment and collection of land taxes
Weakened minimum wage	Employers; the employable	Organised labour	Control over wages paid
Compulsory arbitration	Employers or organised labour[1]	Organised labour or employers[1]	Control over labour disputes
Educational reform	Out-groups	In-groups	Control over curriculum, admissions, staffing and so on

[1] Depending on whom the settlements favour; possibly both will oppose.

Although it includes 11 areas of action, the table is still highly selective; the ILO mission's report shows at least 120 policy areas or specific policies.[1]

These heavy administrative requirements arise at a time when the public services in the developing countries are already over-stretched and over-committed. The execution of a labour-intensive new development policy would thus often be added to already heavy pressures; and it has been described as especially "administration-intensive".[2] One reason for the

[1] Ilchman and Uphoff: "Beyond the economics of labor-intensive development...", op. cit., p. 97, quoting from a paper prepared by Uphoff for an administrative seminar held in Singapore, 1973.

[2] ibid., p. 96.

"administration intensity" of labour-intensive methods in public works, for example, is the time factor: labour-intensive methods will normally take more time (calling as well for more advance planning) and are also assumed to be more subject to delays and disruptions (although this last is doubtful as a general proposition since there can certainly be serious delays and disruptions when capital equipment is deployed). The period of transition to a new technology policy will be especially difficult, since the new technology policies will require arrangements that cut across traditional departmental lines of responsibility. Also, the data basis for the new policies will often be lacking since in the past data collection has been heavily concentrated on the capital-intensive, large-scale sector.

In the case of extension services, links will have to be established with groups (such as informal sector producers) with whom past relations were often more of mutual distrust than of collaboration.

From the administrative viewpoint the implementation of an appropriate technology policy is uncharted territory. There is an urgent need for more research and for arrangements to study and exchange relevant experiences concerning these administrative problems. The administrators, no less than the managers and engineers, will have to change deeply ingrained ways of thinking. In the case of road construction, for example, it has been pointed out in a World Employment Programme study that the administrative procedures (such as conditions of contract, specifications and tendering procedures) used in most developing countries are basically the same as those in the industrialised world.[1] In other words, the administrative procedures were no less "imported" and "inappropriate" than the technology used for the actual road construction.

In so far as a fear of labour problems and labour-management conflicts stands in the way of the adoption of appropriate labour-intensive processes, effective labour legislation and industrial consultation and conciliation methods at the economy and plant level would make an especially important administrative contribution. Here again, as in the case of the presumed greater risks of delay and interruptions connected with labour-intensive methods, real life does not necessarily agree with conventional belief. For example, in one of the World Employment Programme case studies (relating to can making in Thailand) it was suggested that in fact a wide variety of labour-management problems would have been eased considerably if more labour-intensive techniques had been used.[2] For instance, with automated techniques there was friction over inadequate supervision and lack of promotion prospects, and the difficulties of running a high-speed line to produce a variety of products "led to an atmosphere of constant crisis on the factory floor". These are but two examples of oft-repeated hypotheses with

---

[1] Allal and Edmonds: *Manual on the planning of labour-intensive road construction*, op. cit.

[2] See Cooper et al.: "Choice of techniques for can making...", op. cit., p. 115.

administrative implications that seem to call for further testing and fact-finding.

The importance of the social and political setting within which technology policy is placed has repeatedly emerged in the preceding discussion (for instance, Chapter 2 emphasised the different motivations affecting choice of technology on the part of different types of production unit, the special situation regarding choice of technology by multinational enterprises, institutional impediments to the transfer of technology, the importance of an institution such as subcontracting, and so on). Professor Sen, in a study prepared within the framework of the World Employment Programme, has clearly demonstrated the importance of the social setting for the implementation of appropriate technologies. For example, the non-wage modes of production (self-employment, family employment) lend themselves more easily to labour-intensive techniques than do wage modes of employment.[1] Thus, the promotion of appropriate technology requires not only effective action within each mode of production specifically geared to the situation within each mode, but also policies affecting the importance or weight within the economy of the various modes of production. This complicates the tasks of administration but enormously adds to its importance.

Such shifts between sectors and modes of production are the main reason why a capital-intensive technology will not only add less to employment—other things being equal—than a labour-intensive technology, but will often directly *reduce* and *destroy* employment by producing shifts from a labour-intensive non-wage sector (crafts, rural industry, informal small-farm sector) to a wage sector (modern industry, large mechanised farms, and so on). This has been shown by the ILO comprehensive employment strategy mission to Kenya to have been the case for the introduction of a new baking technology, and elsewhere for the production of shoes and sandals. In a wider context it is one of the reasons for the failure of the Green Revolution to have its full potential impact on employment and the reduction of poverty. Conversely, subcontracting, especially the ability to use households on a putting-out basis—so important in Japan's industrial development, for instance—can be a powerful generator of employment, precisely because it shifts production from a wage sector to a non-wage sector, and at the same time from a large-scale sector to a small-scale sector.

TECHNOLOGY PLANNING AND DEVELOPMENT PLANNING

At the first functional level of planning, decision and control, there is a need for close integration with the national plan. As was pointed out in Chapter 1, this has been made easier by new thinking about the purposes of national planning in which the achievement of national technological

---

[1] Sen: *Employment, technology and development,* op. cit., p. 22.

capacity has come to occupy an increasingly important place. Moreover, the adoption of a basic-needs strategy would have a decisive influence on the nature as well as the network of the institutions required and discussed in the earlier part of this chapter.

So technology is both the master and the servant. It is the master because the constraints of technology and technological capacity limit what the planners can reasonably hope to achieve; it is the servant because it has to extend and allocate its resources so as to fit in with the priorities of the planners. Technological capability and the reduction of technological dependence are development objectives in their own right; but at the same time technology policies can be powerfully supported by appropriate direct regulatory measures, such as licensing, rationing of credit or supplies, and so on. The practical conclusion is that staff familiar with science and technology must be involved in the formulation of the national development plan, and that development planners must for their part be involved in the formulation of the science and technology plan.

The science and technology plan usually will involve a much longer time-horizon than the typical four or five years of the national development plan, but this is also true of other elements of national planning, particularly where a new training need or large-scale works are involved (educational plans, irrigation plans, electrification plans, and so forth). Within the technology plan, there will be a mixture of projects which create benefits within a relatively short period, and others which may lead to major breakthroughs only in the long run but whose inclusion is justified by the potential importance of the results in the light of national problems and objectives.

In the inter-relationship between technology policy and development planning two possible chains of causation have been distinguished: *(a)* from a development strategy to project ideas to design and engineering to scientific and technological research; and *(b)* from design and engineering to product and process ideas to project ideas and back to design and engineering. [1] Both these feedbacks are important, and they once again illustrate the mutual relationship of technology and development.

As mentioned in the preceding chapter, the Programme of Action adopted by the World Employment Conference also enjoins developing countries to accelerate the formulation and implementation of a training plan, specifying four levels of training. [2] Although it is not specifically so stated in the Programme of Action, it is clear that such a training plan should harmonise with, and be an integral part of, a long-range technology development plan as well as of over-all manpower planning. All these will be given a special orientation and new focus by a basic-needs strategy and an emphasis on appropriate technology.

---

[1] Bhatt: "On technology policy and its institutional frame", op. cit.
[2] Para. 56.

CONCLUSION

8

In the preceding chapters the discussion of many important issues has already been highly condensed; thus, a full statement of conclusions would be in danger of virtually repeating much of what has been said before. Hence this chapter is limited to a brief statement of conclusions for national action by developing countries and for international action by the world community, as endorsed at the World Employment Conference.

It is now increasingly accepted that the greater emphasis being given to the role, for good or evil, of technology in economic development is going to last and to develop further. Technology will remain one of the key areas of international debate and seems destined to be a prime mover in the hoped-for progress towards a new international economic order. This is in fact shown by the current preparations within the United Nations system for the Conference on Science and Technology for Development, scheduled for 1979.

NATIONAL ACTION BY DEVELOPING COUNTRIES

The creation of strengthened technological capability, separately and jointly, must be one of the primary objectives of developing countries (as was adopted at the Fourth Session of UNCTAD in Nairobi in 1976), both as a development objective in its own right and as an instrument for the reduction of poverty within a basic-needs strategy. The creation and exercise of technological capability involves many different approaches and many different institutions. Hence it requires careful planning and co-ordination, imposing great strains on administrative and planning capacities.

Technological capability includes the better choice and selection from existing technologies available "on the shelf", the adaptation of existing technologies to suit the special needs and requirements of developing countries, the carrying out of applied research and pilot and prototype development as well as, selectively, some basic research on matters of specific and direct concern to developing countries. In the context of a basic-needs strategy, problems of "de-scaling" (that is, production in smaller units and for scattered markets) will be particularly important, as well as the design of new products or adapted products required to satisfy reasonable basic-needs standards. The ultimate objective is to ensure that a technology chosen is both efficient and in accordance with the needs and resources of the country.

The Programme of Action adopted at the World Employment Conference recommended the following types of action at the national level:

52. The choice of appropriate technologies is dependent on the conditions prevailing in each country and the characteristics of each economic sector. This choice must also be based on the full utilisation of national resources. Thus each developing country has the right and duty to choose the technologies which it decides are appropriate. To facilitate such a choice, it will be helpful to establish national subregional and regional centres for the transfer and development of technology and to promote co-operation both between developing countries and between the latter and developed countries. The ILO should help in the establishment of these centres in conjunction with other agencies of the United Nations system.

53. The promotion of research should be a fundamental priority in policies to increase the national technological capacity of developing countries and reduce their dependence on industrialised countries. This research should mainly be undertaken within, and under the direction of, the developing countries themselves or in corresponding regional or subregional bodies where these exist, with the technical and financial assistance of international and other agencies presently involved in such activities. Technological research should furthermore contribute towards the satisfaction of basic needs.

54. Each developing country should accelerate appropriate technological advancement in the informal urban and rural sectors, in particular, to eliminate underemployment and unemployment and raise productivity levels.

55. Foreign firms, in response to the national legislation of developing countries and in negotiation with them, and taking into account the national economic development plans, should:
(a) introduce technologies which are both growth- and employment-generating, directly or indirectly;
(b) adapt technologies to the needs of the host countries, and progressively substitute national for imported technology;
(c) contribute to financing the training of national managers and technicians for the better utilisation and generation of technology;
(d) supply resources and direct technical assistance for national and regional technology research; and
(e) spread technological knowledge and help in its growth by subcontracting the production of parts and materials to national producers, and particularly to small producers.

56. Each developing country should accelerate the formulation and implementation of a training plan at the following levels:
(a) middle-level technicians and skilled workers to be employed in the production technologies associated with the goods and services required to satisfy basic needs;

(*b*) professionals, technicians, managers and skilled workers to replace expatriate staff who presently apply advanced technology;

(*c*) professionals and technicians needed to manage research and studies undertaken by national and/or regional technological research bodies; and

(*d*) technicians, professionals and skilled workers, who should be assured of a measure of social status and incentives to prevent a brain drain, in order to promote the utilisation of technologies designed to achieve material and social objectives.

It may be useful to take into account the above recommendations in the preparation of guidelines for country studies for the United Nations Conference on Science and Technology to be prepared by the member States in consultations with the United Nations system.

INTERNATIONAL ACTION

It is in the interests of the world community as a whole that the efforts of developing countries to acquire greater technological capabilities be supported, and that the present gaps and distortions in the availability of appropriate technologies for different income levels be reduced. It is believed that this is an area where the conditions for international agreement on joint action should be favourable. A number of new proposals were before various international bodies, including the Fourth Session of UNCTAD and the ILO World Employment Conference. Most of these proposals are complementary and indicate that the present international action gap is widely and acutely felt.

The ILO, with its unique tripartite structure, has a significant role to play, in collaboration with other United Nations agencies, to meet the challenge of these tasks. It is in a particularly good position to involve non-governmental organisations and employers' and workers' associations, particularly as regards rural and small-scale technology development and dissemination. In order that the gap between the development of technologies indigenously and their widespread use is narrowed, it is essential that the users (the employers and workers) participate in the decision-making and R and D process at the national, subregional, regional and international levels.

The ILO has in fact been called upon by the World Employment Conference (*a*) to make important contributions to the proposed network for the exchange of technological information, especially as regards appropriate technologies for the rural and urban informal sectors; and (*b*) to reorient and strengthen its existing training programme to provide managerial and vocational training in the choice of appropriate technologies.

The Programme of Action contains the following operative paragraphs relating to action at the international level:

57. International agencies and bilateral and multi-bilateral aid programmes should devote resources and technical assistance to complement developing countries' efforts.

58. At present several organisations of the United Nations system are engaged in work on appropriate technologies for developing countries. Better co-ordination of this work would ensure that the full potential benefits may be realised.

59. The United Nations Inter-Agency Task Force on Information Exchange and the Transfer of Technology is working towards the establishment of a network for the exchange of technological information. At its second session in May 1976 it recommended that:

> organisations of the United Nations system and other organisations having substantive responsibility in the field of technological information and the transfer of technology should develop their relevant activities as components of the over-all network, and in mutual co-operation make available their own information bases and information-handling capabilities as appropriate.

The ILO should strengthen its activities in the field of the collection and dissemination of information on appropriate technologies, especially for the rural sector, and so make an important contribution towards the establishment of the information exchange network referred to above.

60. The ILO should reorient and strengthen its existing programme in order to provide more manpower training and human resources development in the developing countries.

61. The ILO should pursue its research and technical co-operation in the field of development and transfer of technology. It should set up a Working Group in which employers and workers would be represented to examine action on appropriate technology for employment, vocational training and income distribution. The developing countries should participate directly in this Working Group, which should not encroach upon the activities of other United Nations agencies.

62. The Group of 77 endorsed the establishment of a Consultative Group on Appropriate Technology and an International Appropriate Technology Unit, especially directed to research on the choice of alternative use of resources allowing a greater utilisation of labour per unit of investment, provided that such mechanisms are integrated with the ongoing activities of the United Nations system. The Workers' group also endorsed these proposals but emphasised that these bodies should be tripartite in character. Most Western industrialised countries did not support these two proposals. The Workers' group and the Group of 77 supported the UNCTAD proposal for an international code of conduct for technology transfer. This should be of a legally binding, not voluntary, nature. They further supported the suggestion that the Paris Convention of 1883 on industrial property should be drastically revised.

This list indicates the wide range of issues on which new international action is both required and possible. If concerted progress can be made along these lines, an important step will have been taken towards meeting the basic needs of mankind.

APPENDICES

DECLARATION OF PRINCIPLES AND PROGRAMME OF ACTION
adopted by the Tripartite World Conference on Employment, Income Distribution and Social Progress, and the International Division of Labour, Geneva, 4-17 June 1976

A

DECLARATION OF PRINCIPLES

The Tripartite World Conference on Employment, Income Distribution and Social Progress, and the International Division of Labour held in Geneva from 4 to 17 June 1976 in accordance with the resolution adopted by the International Labour Conference during its 59th Session (1974):

AWARE that past development strategies in most developing countries have not led to the eradication of poverty and unemployment; that the historical features of the development processes in these countries have produced an employment structure characterised by a large proportion of the labour force in rural areas with high levels of underemployment and unemployment; that underemployment and poverty in rural and urban informal sectors and open unemployment, especially in urban areas, has reached such critical dimensions that major shifts in development strategies at both national and international levels are urgently needed in order to ensure full employment and an adequate income to every inhabitant of this one world in the shortest possible time;

AWARE that industrialised countries have not been able to maintain full employment and that economic recession has resulted in widespread unemployment;

NOTING that the Conference is a major initiative on the part of the International Labour Organisation towards the efforts that many of the member countries are making to establish a more equitable international economic order, and that it is consistent with the deliberations of the important world conferences of recent years;

RECALLING further the conclusions of the Sixth and Seventh Special Sessions of the United Nations General Assembly, in particular Resolution 3202 (S-VI) concerning the Establishment of a New International Economic Order, and Resolution 3362 (S-VII) concerning Development and International Economic Co-operation;

NOTING that underemployment, unemployment, poverty, malnutrition and illiteracy are caused by both national and international factors; that at the national level they are caused by structural factors emanating from under-development and, at the international level, they are due mainly to the deterio-rating situation in developing countries, which is partly the consequence of cyclical and structural imbalances in the world economic situation;

RECOGNISING that one of the primary objectives of national development efforts and of international economic relations must be to achieve full employ-ment and to satisfy the basic needs of all people throughout this one world;

COMMITTED to the attainment of an equitable distribution of income and wealth through appropriate strategies to eradicate poverty and promote full, productive employment to satisfy basic needs;

NOTING:

(a) that unemployment, underemployment and marginality are a uni-versal concern and affect at least one-third of humanity at the present time, offending human dignity and preventing the exercise of the right to work;

(b) that the experience of the past two decades has shown that rapid growth of gross national product has not automatically reduced poverty and inequality in many countries, nor has it provided sufficient productive employ-ment within acceptable periods of time;

(c) the current unsatisfactory international economic situation and the discussions of problems affecting unemployment and related issues in UNCTAD IV;

(d) that the existence of an informal urban sector which has grown out of proportion during the past decades in the developing countries and the chronic lack of jobs in rural areas burden the labour markets and hinder the sectoral and regional integration of national development policies;

(e) that it is necessary to replace the current international division of labour wherein the participation of developing countries in international trade is mainly the exportation of raw materials, semi-processed products and highly labour-intensive manufactured goods and the importation of highly capital-intensive industrial products, so as to enable all countries to engage in other types of production in accordance with their national priorities;

RECALLING the Universal Declaration of Human Rights, in particular Article 23, adopted by the General Assembly of the United Nations in 1948;

CONSIDERING that only productive work and gainful employment, without discrimination, enable man to fulfil himself socially and as an individual, and reconfirming that the assured opportunity to work is a basic human right and freedom;

CONSIDERING that the growth of productive employment is one of the most effective means to ensure a just and equitable distribution of income and to raise the standard of living of the majority of the population;

CONVINCED that the establishment and modernisation of small and medium-sized enterprises in rural as well as in urban sectors will increase the volume of employment and therefore play an important part in a basic-needs strategy, and that the private sector has an important role to play in development and employment creation;

CONSIDERING that integrated development of developing countries can be achieved only in so far as equal priority is attached to the social, economic and political aspects of development;

AFFIRMING that the problems of underemployment, unemployment and poverty must be attacked by means of direct, well co-ordinated measures at both national and international levels;

RECOGNISING that in most developing countries, the government is the principal promoter of development and employment and the competent instrument to achieve a just and equitable distribution of income, with the effective participation of trade unions, rural workers' organisations and employers' associations;

RECOGNISING that international relations should be based on co-operation, interdependence, national sovereignty, self-determination of peoples, and non-intervention in the internal affairs of countries;

RECONFIRMING the importance of regional and subregional co-operation as a major instrument to achieve the expansion of domestic markets, to facilitate the use of modern technologies, efficient industrialisation, better integration into the world economy, and to give greater weight to the positions of developing countries in international relations, with a view to accelerating the development of Third World countries;

NOTING the firm commitment of the developing countries and of some developed countries to implement the New International Economic Order, based on the principles contained in the Charter of Economic Rights and Duties of States;

NOTING that a review and appraisal of the strategy for the Second Development Decade (Resolution 3517 of the United Nations General Assembly) are taking place and that preparations for the Third Development Decade have commenced;

CONVINCED that the strategy for the Second Development Decade needs to be complemented by a programme of action to guide international and national development efforts towards fulfilling the basic needs of all the people and particularly the elementary needs of the lowest income groups;

RECALLING that the ILO, particularly through its World Employment Programme, has a direct responsibility for elaborating such a strategy with regard to the achievement of full productive employment in decent working conditions, and ensuring respect for the freedoms and rights of association and collective bargaining laid down in Conventions Nos. 87, 98 and 135;

The Conference hereby adopts this Declaration of Principles and the Programme of Action and requests the Governing Body of the ILO to implement the Programme of Action where appropriate in co-operation with other international organisations.

PROGRAMME OF ACTION

I. Basic needs

1. Strategies and national development plans and policies should include explicitly as a priority objective the promotion of employment and the satisfaction of the basic needs of each country's population.

2. Basic needs, as understood in this Programme of Action, include two elements. First, they include certain minimum requirements of a family for private consumption: adequate food, shelter and clothing, as well as certain household equipment and furniture. Second, they include essential services provided by and for the community at large, such as safe drinking water, sanitation, public transport and health, educational and cultural facilities.

3. A basic-needs-oriented policy implies the participation of the people in making the decisions which affect them through organisations of their own choice.

4. In all countries freely chosen employment enters into a basic-needs policy both as a means and as an end. Employment yields an output. It provides an income to the employed, and gives the individual a feeling of self-respect, dignity and of being a worthy member of society.

5. It is important to recognise that the concept of basic needs is a country-specific and dynamic concept. The concept of basic needs should be placed within a context of a nation's over-all economic and social development. In no circumstances should it be taken to mean merely the minimum necessary for subsistence; it should be placed within a context of national independence, the dignity of individual and peoples and their freedom to chart their destiny without hindrance.

Strategies and policies to create full employment
and to meet basic needs in developing countries

6. In developing countries satisfaction of basic needs cannot be achieved without both acceleration in their economic growth and measures aimed at changing the pattern of growth and access to the use of productive resources by the lowest income groups. Often these measures will require a transformation of social structures, including an initial redistribution of assets, especially land, with adequate and timely compensation. Land reform should be supplemented by rural community development. In some countries, however, public ownership and control of other assets is an essential ingredient of their strategy. Obviously, each country must democratically and independently decide its policies in accordance with its needs and objectives.

7. Any national employment-centred development strategy aiming at satisfying the basic needs of the population as a whole should, however, include the following essential elements, to the extent that countries consider them to be desirable:

Macro-economic policies

(a) An increase in the volume and productivity of work in order to increase the incomes of the lowest income groups;

(b) strengthening the production and distribution system of essential goods and services to correspond with the new pattern of demand;

(c) an increase in resource mobilisation for investment; the introduction of progressive income and wealth taxation policies; the adoption of credit policies to ensure employment creation and increased production of basic goods and services;

(d) the control of the utilisation and processing of natural resources as well as the establishment of basic industries that would generate self-reliant and harmonious economic development;

(e) developing inter-regional trade, especially among the developing countries, in order to promote collective self-reliance and to ensure the satisfaction of basic import needs without depending permanently on external aid;

(f) a planned increase in investments in order to achieve diversification of employment and technological progress and to overcome other regional and sectoral inequalities;

(g) reform of the price mechanism in order to achieve greater equity and efficiency in resource allocation and to ensure sufficient income to small producers;

(h) reform of the fiscal system to provide employment-linked incentives and more socially just patterns of income distribution;

(i) safeguarding ecological and environmental balances;

(j) provision by the government of the policy framework to guide the private and public sectors towards meeting basic needs, and making its own industrial enterprises model employers; in many cases this can only be done in a national planning framework;

(k) the development of human resources through education and vocational training.

Employment policy

8. Member States should place prime emphasis on the generation of employment, in particular to meet the challenge of creating sufficient jobs in developing countries by the year 2000 and thereby achieve full employment. Specific targets should be set to reduce progressively unemployment and under-employment.

9. The following policies should be adopted to encourage employment creation:

(a) Member States should ratify ILO Convention No. 122 and should ratify, implement and safeguard fair labour standards, such as the right to organise and to engage in collective bargaining, as laid down in ILO Conventions Nos. 87, 98 and 135.

(b) In the criteria for project selection and appraisal, employment and income distribution aspects should have adequate emphasis in development planning and in the lending policies of international financial institutions.

(c) Member States should implement active labour market policies of the type set forth in the ILO Human Resources Development Convention, 1975 (No. 142), and the accompanying Human Resources Development Recommendation, 1975 (No. 150), and adjust enterprise-level policies, especially with regard to recruitment, work organisation, working conditions and work content, so as fully to absorb underutilised labour resources.

(d) Wage policies should be such that:

(i) they ensure minimum levels of living;

(ii) the real wages of workers and the real incomes of self-employed producers are protected and progressively increased;

(iii) wage levels are equitable and reflect relative social productivity;

(iv) anti-inflationary incomes and price policies, where introduced, take these objectives into account.

(e) Equality of treatment and remuneration for women should be ensured.

Rural sector policies

10. Governments should give high priority to rural development, and increase the effectiveness of their policies, including those to reorganise the agrarian structure. Rural development involves the modernisation of agriculture,

the development of agro-based industries, and the provision of both physical and social infrastructure. It should encompass educational and vocational training facilities, the construction of main and feeder roads, the provision of credit facilities and technical assistance, especially to small farmers and agricultural labourers.

11. Co-operatives should be promoted in accordance with ILO Recommendation No. 127 and extend not only to the use of land, equipment and credit, but also to the fields of transportation, storage, marketing and the distribution network, processing and services generally. More emphasis should be placed on the development of co-operatives in national policies, especially when they can be implemented so as to involve the lowest income groups, through their own organisations.

12. In most of the developing countries, agrarian reform, land distribution and the provision of ancillary services are basic to rural development. A minimum requirement is to provide house sites for rural and plantation workers and other landless labourers so as to assist them in building their homes and making them independent, especially in case of loss of employment.

13. The main thrust of a basic-needs strategy must be to ensure that there is effective mass participation of the rural population in the political process in order to safeguard their interests. In view of the highly hierarchical social and economic structure of agrarian societies in some developing countries, measures of redistributive justice are likely to be thwarted unless backed by organisations of rural workers. A policy of active encouragement to small farmers and rural workers' organisations should be pursued to enable them to participate effectively in the implementation of:

(a) programmes of agrarian reforms, distribution of surplus lands and land settlement;

(b) programmes for developing ancillary services such as credit, supply of inputs and marketing; and

(c) programmes concerning other employment generation schemes, such as public works, agro-industries and rural crafts.

As specified in ILO Convention No. 141, governments should create conditions for the development of effective organisations of rural workers.

Social policies

14. Social policies should be designed to increase the welfare of working people, especially women, the young and the aged.

Women

15. Since women constitute the group on the bottom of the ladder in many developing countries in respect of employment, poverty, education,

training and status, the Conference recommends that special emphasis be placed in developing countries on promoting the status, education, development and employment of women and on integrating women into the economic and civic life of the country.

16. Specifically, the Conference recommends:

(a) the abolition of every kind of discrimination as regards the right to work, pay, employment, vocational guidance and training (including in-service training), promotion in employment and access to skilled jobs;

(b) that more favourable working conditions be ensured so that women may perform their other functions in society and married women may be able to return to either full-time or part-time productive employment;

(c) that the work burden and drudgery of women be relieved by improving their working and living conditions and by providing more resources for investment in favour of women in rural areas.

The young, the aged and the handicapped

17. In the implementation of basic-needs strategies, there should be no discrimination against the young, the aged or the handicapped. Every effort should be made to provide the young with productive employment, equal opportunity and equal pay for work of equal value, vocational training and working conditions suited to their age. Exploitation of child labour should be prohibited in accordance with the relevant ILO standards.

Participation of organised groups

18. Governments must try to involve employers' organisations, trade unions and rural workers' and producers' organisations in decision-making procedures and in the process of implementation at all levels. These are the organisations which represent the vast majority of the population and, therefore, they must be the ones to help define the basic needs and apply the necessary strategies.

19. Employers' and producers' organisations, trade unions and other workers' organisations such as rural workers' organisations have an important role to play in the design and implementation of sucessful development strategies. They must be encouraged to participate effectively in the decision-making process. Workers' organisations are also of great importance in the search for a reform of the existing international economic structures and they have a major role to play in the achievement of a fairer distribution of income and wealth.

Education

20. Education is itself a basic need, and equality of access to educational services, particularly in rural areas, is therefore an important ingredient of a

basic-needs strategy. Lack of access to education denies many people, and particularly women, the opportunity to participate fully and meaningfully in the social, economic, cultural and political life of the community.

21. Educational and vocational training systems should be adapted to national development needs and should avoid an élitist bias; priority should be given to adult and primary education, especially in the rural areas.

Population policy

22. High birth rates in poverty-stricken areas are not the cause of under-development but a result of it. They may, however, jeopardise the satisfaction of basic needs. It is only through the fulfilment of these needs, with special emphasis on the development of the position and status of women, that couples will be in a better position to determine the size of their family in a manner compatible with the aims of their society. The Conference is of the view that population policies consistent with the culture and the societies involved, as recommended by the 1974 World Population Conference, should be strongly encouraged. It recommends that information on population programmes should be made available to people in a form and language that they can understand.

International economic co-operation

23. The satisfaction of basic needs is a national endeavour, but its success depends crucially upon strengthening world peace and disarmament and the establishment of a New International Economic Order. The World Employ-ment Conference fully supports the efforts being made by the United Nations General Assembly in its resolution as adopted at the Seventh Special Session and through the relevant agencies of the UN system to introduce international reforms in trade and finance in favour of developing countries and thus to contribute to the creation of a New International Economic Order. The Con-ference recognises that the basic-needs strategy is only the first phase of the redistributive global growth process.

24. In particular, the Conference, recognising the primary objectives of national development, urges ILO member States to continue their efforts through the appropriate UN agencies to:

(a) stabilise developing countries' exports of primary products and improve their terms of trade through financing an integrated commodity programme;

(b) secure expanded access for developing countries' manufactured exports to the markets of rich countries through trade liberalisation measures on a non-reciprocal basis;

(c) increase the net transfer of resources to developing countries, includ-ing the mitigation of their debt burden;

(d) increase mutual economic co-operation between countries with different social and economic systems.

25. The Employers' group wished it to be placed on record that they regarded the section on international economic co-operation as being outside the proper competence of the ILO and as being inappropriate for comment by employers. They agreed, however, that the ILO should co-operate with relevant UN organisations, wherever appropriate, in implementing its policies throughout the developing world.

26. A number of Western industrialised countries wished it to be placed on record that they regarded the section on international economic co-operation (paras. 23 and 24) as being outside the proper competence of the ILO. They took the view that, within its area of competence, the ILO should co-operate with other UN organisations, wherever appropriate, in implementing its policies throughout the developing world.

Recommendations

27. The ILO should co-operate with other UN agencies in bringing about these desired reforms in order to give meaning and reality to the expressed commitment of the world community to assist national basic-needs strategies. It should work through, in particular, the World Employment Programme, including its regional components, and its recognised instrumentalities, such as the existing standard-setting activities, technical assistance and industrial activities.

28. The ILO should, in particular, undertake promotion of short-term and quick employment-generating programmes for making an immediate impact on the prevailing levels of poverty and massive waste of human resources. The Conference recommends that a portion of the $1,000 million International Fund for Agricultural Development should be used for the generation of employment in the rural sector.

29. The Governing Body of the ILO is urged to recommend the review of research programmes, operational activities and organisational structures of the UN family so as to focus them more sharply on the contribution they can make to meeting the basic-needs targets, particularly of the lowest income groups. The Administrative Committee on Co-ordination (ACC) should be requested to review, monitor and report on the work of the different agencies and regional commissions of the UN system.

30. The ILO should, in co-operation both with other UN bodies and with interested national governments, consider the feasibility of initiating a world-wide programme in support of household surveys to map the nature, extent and causes of poverty; to assist countries to set up the necessary statistical and monitoring services; and to measure progress toward the fulfilment of basic needs.

31. Member States should, to the extent possible, supply the ILO, before the end of the decade, with the following information:

(a) a quantitative evaluation of basic needs for the lowest income groups within their population, preferably based on the findings of a tripartite commission established for the purpose;

(b) a description of policies, existing and in preparation, in order to implement the basic-needs strategy.

32. The ILO is requested to prepare a report for an annual conference before the end of the decade and to include the following information:

(a) an elaboration of more precise concepts defining basic needs on the basis of national replies;

(b) a survey of the entire range of national replies received and an analysis of the national situations with respect to the levels of basic needs as well as policies to attain them.

33. The Governing Body of the ILO is urged to place the question of the revision of Convention No. 122 on the agenda of an early session of the International Labour Conference.

34. The Conference finally requests that policies required to meet basic needs become an essential part of the Second United Nations Development Decade Strategy and form the core of the Third Development Decade Strategy.

II. International manpower movements and employment

General objectives of national and international policies

35. The aim of national and international policies in this field should be threefold: (i) to provide more attractive alternatives to migration in the country of origin; (ii) to protect migrants and their families from the difficulties and distress which sometimes follow migration; (iii) to take care that neither migration nor its alternatives are prejudicial to the rest of the population or harmful to economic and social development in either the country of origin or the country of employment.

Measures designed to avoid the need for workers to emigrate

36. The development strategy in the countries of origin should include in particular an employment policy which would give workers productive employment and satisfactory conditions of work and life.

37. This strategy should be implemented in the framework of multilateral and bilateral co-operation which would make it possible through such means as encouragement of appropriate intensified capital movements and transfers of technical knowledge to promote a reciprocally advantageous international

division of labour; this calls for necessary readjustments in countries of employment.

Measures against migrations in abusive conditions and in favour of the promotion of equality of opportunity and treatment

38. Governments, employers and workers of the countries of employment should ensure that all migrants are protected against any exploitation and effectively enjoy equality of opportunity and treatment. These principles and the means of implementing them are stated in detail in the international standards of the ILO and more specifically in the Migrant Workers (Supplementary Provisions) Convention, 1975 (No. 143), and in the complementary Migrant Workers Recommendation, 1975 (No. 151). A special effort should be made to ratify and apply the Convention and give effect to the provisions of the Recommendation, especially with regard to:

(a) the fight against migrations in abusive conditions, particularly through sanctions in conformity with Article 6 of the Convention;

(b) the promotion of equality of opportunity and treatment in respect of employment and occupation, of social security, of trade union and cultural rights and of individual and collective freedoms and especially the encouragement of the efforts of migrant workers and their families to preserve their national and ethnic identity as well as their cultural ties with their country of origin, including the possibility for children to be given some knowledge of their mother tongue;

(c) the elaboration and implementation of a social policy which emphasises:

  (i)   reunification of families;
  (ii) protection of the health of migrant workers;
  (iii) establishment of adequate social services;
  (d) minimum guarantees as regards employment and residence.

39. In order to combat discrimination and illegal trafficking in manpower, governments, employers and workers should strengthen their action to ensure the application of national legislation and collective agreements and to initiate the early introduction of appropriate penal sanctions against all who organise or knowingly take advantage of illegal movements of manpower.

Multilateral and bilateral agreements

40. Multilateral and bilateral agreements should be drawn up to deal with the migration of workers and problems concerning migrant workers and their families. Such agreements should be in accordance with the principles established in ILO standards. As far as possible, representative organisations of employers and workers should participate in their preparation and implementation.

41. Such agreements should be based upon the economic and social needs of the countries of origin and the countries of employment; they should take account not only of short-term manpower needs and resources, but also of the long-term social and economic consequences of migration, for migrants as well as for the communities concerned.

42. One of the principal objectives of mutually accepted policies in the framework of these agreements should be to even out fluctuations in migration movements, return migration flows and remittances and make them as far as possible predictable, continuous and assured so as to facilitate the implementation of long-term programmes of economic and social development.

43. Taking into account the economic and social circumstances in the countries and regions concerned and the characteristics of the migration movements concerned, these agreements should in appropriate cases:

(a) facilitate the co-ordination of employment policies, especially in the framework of efforts for economic and social integration on a regional basis;

(b) regulate the recruitment of migrant workers without discrimination and according to their free choice under the auspices of the employment services of the countries concerned;

(c) provide for periodic exchange of information between the countries concerned on the occupational categories and the number of workers to whom contracts could be offered and who would be ready to emigrate or return to their country of origin; for this purpose skilled manpower pools or data banks should be established to provide reliable information on the supply of and demand for skilled, professional and technical manpower;

(d) reinforce co-operation between the employment and other services dealing with migration and migrant workers in the countries concerned;

(e) give priority to the recruitment of workers who are underemployed or unemployed;

(f) provide that countries of origin should adopt appropriate measures so as to avoid the departure of skilled workers, including adaptation of education and training schemes to national needs and offering highly trained personnel conditions permitting them to remain and serve their own country;

(g) provide that countries of employment should refrain from recruiting skilled and highly skilled workers when there are recognised or potential shortages of such workers in the country of origin;

(h) provide that the countries of employment could take complementary measures to aid the developing countries to minimise their loss of qualified manpower, for example by increasing training possibilities for their own nationals in those fields where skills are scarce and by eliminating any part of their immigration laws and regulations which have the effect of encouraging the entry of professional and other highly qualified migrants;

(i) provide ways of limiting losses in countries of origin, particularly developing countries, which may result from the departure of skilled personnel whose education and training they have provided;

(j) establish training facilities, where these do not already exist, making possible:

(i) suitable preparation, documentation and training of candidates for emigration;

(ii) vocational training and advancement of migrant workers in the country of employment;

(iii) training of workers wishing to return to the countries of origin, taking account of the aptitudes of such workers and the needs of their countries;

(k) adopt the necessary measures to facilitate the voluntary return to their countries of origin of migrant workers and their resettlement;

(l) provide for social security benefits for families which have stayed in the country of origin and for suitable means of ensuring that migrant workers returning to their home countries enjoy continuity of social security benefits;

(m) take into consideration the need for financing the above measures by appropriate means.

The role of the ILO

44. At their request, the ILO should provide technical co-operation to the countries concerned and technical support to regional organisations in order to make it possible to prepare and implement the above measures.

45. At the request of governments concerned, the ILO should study the possibility of setting up at regional or subregional level a system designed with the collaboration of the representative employers' and workers' organisations concerned to improve information on the availability of job opportunities in certain industries and certain types of employment for the benefit of candidates for emigration or return to their country of origin.

46. The Office should:

(a) initiate studies on the economic and social effects of different kinds of migration for employment;

(b) make studies and organise meetings at regional or subregional levels on the problems of migrant workers who have not been regularly admitted or who lack official papers.

III. Technologies for productive employment creation in developing countries

Policy objectives

47. Technology has an important role to play in the process of development. Since technology is linked with the choice of products as well as with capital investment, labour and skills required to produce them, it has a bearing on the level of productive employment and the distribution of income. Technology, therefore, is an important element of the basic-needs strategy, which must be part of an over-all national, economic and social development strategy.

48. There is an urgent need for appropriate and optimal technology, that is, management and production techniques which are best suited to the resources and future development potential of developing countries. Such technology should contribute to greater productive employment opportunities, elimination of poverty and the achievement of equitable income distribution.

49. The exclusive use of labour-intensive techniques will neither solve the problems of the developing countries nor reduce their dependence on industrialised countries. Likewise, the exclusive use of capital-intensive techniques will present the developing countries with serious problems: financial difficulties, lack of managerial staff and supervisory personnel and delays in the solution to employment problems. Thus developing countries should arrive at a reasonable balance between labour-intensive and capital-intensive techniques, with a view to achieving the fundamental aim of maximising growth and employment and satisfying basic needs. This strategy of equilibrium between the various types of technologies should also take account of the desire to adopt advanced techniques, with a view to reducing the existing technological gap between countries.

50. Choice, development and transfer of technology require that proper emphasis should be placed on the building up of national infrastructure for human resources development, particularly to promote training of workers, technicians and managers for appropriate technology selection.

51. In the selection of new technologies appropriate to their needs, the developing countries should take due account of the need to protect their ecology and natural resources. There is also a need to pay due attention to social aspects, working conditions and the safety of workers when introducing new technologies.

Action at the national level

52. The choice of appropriate technologies is dependent on the conditions prevailing in each country and the characteristics of each economic sector. This choice must also be based on the full utilisation of national resources. Thus each developing country has the right and duty to choose the technologies

which it decides are appropriate. To facilitate such a choice, it will be helpful to establish national subregional and regional centres for the transfer and development of technology and to promote co-operation both between developing countries and between the latter and developed countries. The ILO should help in the establishment of these centres in conjunction with other agencies of the UN system.

53. The promotion of research should be a fundamental priority in policies to increase the national technological capacity of developing countries and reduce their dependence on industrialised countries. This research should mainly be undertaken within, and under the direction of, the developing countries themselves or in corresponding regional or subregional bodies where these exist, with the technical and financial assistance of international and other agencies presently involved in such activities. Technological research should furthermore contribute towards the satisfaction of basic needs.

54. Each developing country should accelerate appropriate technological advancement in the informal urban and rural sectors, in particular, to eliminate underemployment and unemployment and raise productivity levels.

55. Foreign firms, in response to the national legislation of developing countries and in negotiation with them, and taking into account the national economic development plans, should:

(a) introduce technologies which are both growth- and employment-generating, directly or indirectly;

(b) adapt technologies to the needs of the host countries, and progressively substitute national for imported technology;

(c) contribute to financing the training of national managers and technicians for the better utilisation and generation of technology;

(d) supply resources and direct technical assistance for national and regional technology research; and

(e) spread technological knowledge and help in its growth by subcontracting the production of parts and materials to national producers, and particularly to small producers.

56. Each developing country should accelerate the formulation and implementation of a training plan at the following levels:

(a) middle-level technicians and skilled workers to be employed in the production technologies associated with the goods and services required to satisfy basic needs;

(b) professionals, technicians, managers and skilled workers to replace expatriate staff who presently apply advanced technology;

(c) professionals and technicians needed to manage research and studies undertaken by national and/or regional technological research bodies; and *(d)* technicians, professionals and skilled workers, who should be assured of a measure of social status and incentives to prevent a brain drain, in order to promote the utilisation of technologies designed to achieve material and social objectives.

Action at the international level

57. International agencies and bilateral and multi-bilateral aid programmes should devote resources and technical assistance to complement developing countries' efforts.

58. At present several organisations of the UN system are engaged in work on appropriate technologies for developing countries. Better co-ordination of this work would ensure that the full potential benefits may be realised.

59. The UN Inter-Agency Task Force on Information Exchange and the Transfer of Technology is working towards the establishment of a network for the exchange of technological information. At its second session in May 1976 it recommended that:

> organisations of the United Nations system and other organisations having substantive responsibility in the field of technological information and the transfer of technology should develop their relevant activities as components of the over-all network, and in mutual co-operation make available their own information bases and information-handling capabilities as appropriate.

The ILO should strengthen its activities in the field of the collection and dissemination of information on appropriate technologies, especially for the rural sector, and so make an important contribution towards the establishment of the information exchange network referred to above.

60. The ILO should reorient and strengthen its existing programme in order to provide more manpower training and human resources development in the developing countries.

61. The ILO should pursue its research and technical co-operation in the field of development and transfer of technology. It should set up a Working Group in which employers and workers would be represented to examine action on appropriate technology for employment, vocational training and income distribution. The developing countries should participate directly in this Working Group, which should not encroach upon the activities of other UN agencies.

62. The Group of 77 endorsed the establishment of a Consultative Group on Appropriate Technology and an International Appropriate Technology Unit, especially directed to research on the choice of alternative use

of resources allowing a greater utilisation of labour per unit of investment, provided that such mechanisms are integrated with the ongoing activities of the UN system. The Workers' group also endorsed these proposals but emphasised that these bodies should be tripartite in character. Most Western industrialised countries did not support these two proposals. The Workers' group and the Group of 77 supported the UNCTAD proposal for an international code of conduct for technology transfer. This should be of a legally binding, not voluntary, nature. They further supported the suggestion that the Paris Convention of 1883 on industrial property should be drastically revised.

IV. Active manpower policies and adjustment assistance in developed countries

General principles

63. Governments of developed countries should pursue a determined policy to achieve and maintain full employment, i.e. to provide employment opportunities for all those who want to work, and contribute to a fair distribution of income and wealth in these countries. Employment policy should be closely integrated with over-all economic policy and national planning. It has to be related to other social policies.

64. The success of active manpower policies pursued with this aim will facilitate adaptation to structural changes including those which result from expanding trade with developing countries, thereby supporting growth and increased employment in these countries. Employment policy should not exclusively be based on measures to influence general demand. It should also be based on a range of selective measures to create new job opportunities. Such selective measures should also make a contribution to the struggle against inflation. Governments in the industrialised countries should strengthen the co-ordination of economic policies to maintain full employment. Measures should also be taken to ensure close collaboration concerning the migratory movement of workers between countries of origin and reception.

65. This policy will contribute to a high level of economic activity and improvements in the international economic order as called for by the UN General Assembly, and will lead to increased trade with the developing countries, thus increasing growth and employment in these countries.

66. Structural changes resulting from modifications in the international economic order must not take place at the expense of workers. Such changes should contribute to job creation in both industrialised and developing countries, and assure suitable employment to all workers, involving countries of whatever social and political system. The governments concerned should provide adjustment assistance in order to facilitate the establishment of new economic relations between developing and developed nations. It is envisaged that such adjustment assistance would not diminish development aid.

Policy measures

67. The priorities of national employment policies should be:

(i) the maintenance of as high a demand for labour as is necessary in order to achieve full employment;

(ii) measures and policies to promote stable economic growth, which should include both general and selective measures;

(iii) the reinforcement of measures designed to provide protection against undesirable effects of cyclical evolution or structural change, such as those mentioned in ILO Conventions and Recommendations.

These measures could include:

— provision of maximum practicable notice of change for workers whose jobs are threatened;

— provision of appropriate income levels for a reasonable period and the safeguarding of pension rights;

— provision of retraining;

— provision of special measures for women, migrants, young workers, and handicapped workers whose re-employment involves special problems.

These matters should be dealt with in close co-operation between governments, employers and workers.

68. Many of these features already exist in the policies of industrialised countries.

69. In implementing employment and manpower policies, the industrialised countries should continue to pursue and expand trade liberalisation policies in order to increase imports of manufactures and semi-manufactures from developing countries in an effort to increase their employment and incomes, while continuing to maintain employment in industrialised countries. Adjustment assistance is considered preferable to import restrictions.

70. Consistent with national laws and systems, adjustment assistance should start well before workers lose their jobs, when this can be clearly established, and not only when unemployment is imminent.

71. Regional or national readjustment funds could be set up by the industrialised countries or existing funds (for example the EEC Social and Regional Funds) could be adapted for the purpose of assisting in the adjustment of industries and workers affected by changes in the international economic situation. This ought not to reduce development aid.

72. The competitiveness of new imports from developing countries should not be achieved to the detriment of fair labour standards.

73. The World Employment Conference expresses the hope that the discussions in the Multilateral Trade Negotiations concerning the GATT safeguard clause, i.e. GATT Article 19, will lead to improvements in the international safeguard system.

74. Governments and employers' and workers' organisations shall work together to improve industrial life. Employers and workers should consider participation by workers in matters of recognised mutual concern.

Proposals for an ILO action programme

75. The traditional role of the ILO regarding labour standards should be continued in order to ensure respect for fair labour standards in developing and industrialised countries alike.

76. The ILO could contribute to the exchange of information and experience on the functioning and problems of active manpower policies and adjustment assistance. The Workers' members felt that the ILO could, within its special competence and in the context of multilateral trade negotiations, contribute to the improvement of an international safeguard system covering employment and income guarantees, fair labour standards and adjustment measures.

77. ILO Industrial Committees could provide a forum for discussing the problems of employment and working conditions resulting from structural change.

78. The Turin Centre, CINTERFOR and other regional vocational training centres have an essential role to play in training, a role which could usefully be widened into areas not currently covered.

V. The role of multinational enterprises in employment creation in the developing countries

The Conference was unable to reach a consensus on the role of multinational enterprises in developing countries. The following paragraphs reflect the position of the different parties.

Declarations of Government members

79. Some governments stressed the positive aspects of the activities of multinational enterprises in developing countries, which they saw as direct employment creation, the linkage effects on the economy, the firms' contribution to an improvement of training, the creation of social services, etc.

80. Some governments stressed that multinational corporations had a role to play in the implementation of a basic-needs strategy. However, it is necessary first to identify the different types of corporation according to their objectives in order to determine which ones could be expected to contribute to the implementation of a basic-needs strategy.

81. Some governments on the other hand underlined the negative effects of the activities of multinational corporations in developing countries, which they saw as the creation of an international division of labour unfavourable to these countries, the control of raw materials, the lack of respect for the sovereign rights of States, the insecurity of the employment provided, the lack of respect for trade union rights and notably the expatriation of profits.

82. Some governments felt that efforts should be made to try to reinforce co-operation between host countries and multinational enterprises, especially through the creation of a favourable climate for private foreign investments. In addition, according to these governments, multinational corporations should not be treated less favourably than local companies.

83. Other governments expressed the opinion that the application of discriminatory measures with regard to multinational enterprises as opposed to local enterprises was one of the sovereign rights of States.

84. The Government members of countries belonging to the Group of 77 based their position on Resolution 3201 adopted by the General Assembly of the United Nations on 1 April 1974 on the establishment of a New International Economic Order based on equity, equality, sovereign rights, interdependence, common interests and co-operation between all States regardless of their economic and social systems, as well as on the conclusions and recommendations adopted by the Fourth Conference of Non-Aligned Countries in Algiers. These countries stated that transnational enterprises were responsible for the world-wide economic imbalance, that they infringed the sovereignty of States, and that they sometimes tended to constitute monopolies and to engage in market sharing and fixing prices. These governments maintained that all action vis-à-vis transnational enterprises must be taken within the framework of a global strategy conceived to bring about quantitative and qualitative changes in the present system of economic and financial relations. They recalled the sovereign rights of States and condemned all interference in the internal matters of the countries in which transnational enterprises invested.

85. The member countries of the Group of 77 recommended strengthening national enterprises to enable them to take necessary steps with a view to preventing the negative effects of the activities of transnational corporations (TNCs). They also recommended that member States and the ILO continue to provide full support to the activities of the UN Commission on Transnational Corporations to regulate the activities of such enterprises particularly in relation to the Code of Conduct which TNCs should observe, containing the following basic principles:

(i)   TNCs must be subject to the laws and regulations of the host country and in the event of a dispute accept the exclusive jurisdiction of the courts of the country in which they operate;

(ii)   TNCs should refrain from all interference in the internal affairs of the States in which they operate;

(iii)   TNCs should refrain from interference in their relations between the government of the host country and other States, and from influencing these relations;

(iv)   TNCs should not serve as an instrument of the external policy of another State nor as a means of extending to the host country juridical regulations of the country of origin;

(v)   TNCs should be subject to the permanent sovereignty which the host country exercises over all its wealth, natural resources and economic activities;

(vi)   TNCs should comply with national development policies, objectives and priorities and make a contribution to their implementation;

(vii)   TNCs should supply the government of the host country with relevant information on their activities in order to ensure that those activities are in accordance with the national development policies, objectives and priorities of the host country;

(viii) TNCs should conduct their operations in such a way that they result in a net inflow of financial resources for the host country;

(ix)   TNCs should contribute to the development of the domestic, scientific and technological capacity of the host countries;

(x)   TNCs should refrain from restrictive trade practices;

(xi)   TNCs should respect the socio-cultural identity of the host country.

86. The Group of 77 also recommended that developing countries adopt measures at the national, regional and international levels in order to ensure that transnational enterprises should reorient their activities so as to undertake further manufacturing processes in developing countries and processing in those countries of raw materials for national or foreign markets. They also recommended that the ILO and member States co-operate with a view to bringing the UN Commission on Transnational Corporations to consider among the points to be included in the compulsory Code of Conduct of TNCs those concerning the obligation of these enterprises to hire local labour, not to discriminate against local workers in respect of salaries, conditions of work, training, promotion and access to different levels of seniority. And lastly they recommended that developing countries take steps in order to regulate and control the activities of transnational enterprises so as to ensure that they would act as a positive factor supporting the efforts of developing countries to expand their exports, through the direct impact which the diversification

and expansion of such exports can have on the generation of productive employment.

87. The Group of 77 considered that, in conformity with the policies laid down in national development plans, and adhering to the national laws and priorities, and fully respecting the sovereignty of the host countries, the transnational corporations should:

(i) introduce technologies which are both growth- and employment-generating, directly or indirectly;

(ii) adapt technologies to the needs of the host countries;

(iii) contribute to financing the training of national managers and technicians for the better utilisation of technology;

(iv) supply resources and direct technical assistance for national and regional technology research;

(v) spread technological knowledge and help in its growth by sub-contracting the production of parts and materials to national producers and particularly to small producers;

(vi) disclose and fully make available to the host countries all the technical know-how and information involved in production maintenance, design construction, research and development, etc.

88. The Group of 77 supported the proposals of the Workers' group set out in paragraph 118 (i)—(v) below, in particular the suggestion that the ILO Governing Body should place the issue of transnational enterprises and social policy on the agenda of the 1978 Session of the International Labour Conference in order that Conventions on TNCs should be adopted in the following areas: industrial relations, employment and training, conditions of life and work.

89. Government members of the European socialist countries supported in principle the position of the Group of 77 as well as that of the Workers' members, and endorsed the proposal to place the issue of multinational enterprises and social policy on the agenda of the International Labour Conference in 1978. They felt that in the countries where multinational enterprises operated, they should contribute to employment creation without hindering either a just distribution of incomes or social progress. They underlined that States had an unconditional right to control the activities of multinational enterprises, and that these enterprises must respect the sovereign rights of States and must not interfere in their internal affairs.

90. Most Government members of industrialised market economy countries underlined the positive effects of the activities of multinational enterprises on the economic development of developing countries. These governments underlined the importance of the task of all countries concerned in assisting the economic development of the Third World. They were of the opinion

that the multinational enterprises could contribute to the economic development of the host country, especially through the creation of employment. The governments of home countries of multinational enterprises, while considering their own national requirements, should continue to apply selective incentives for foreign investments in such a way as to encourage investments which met the basic needs of the host country. Countries which welcomed foreign investment should create a favourable and stable investment climate which encouraged multinational enterprises to adapt their activities to the economic needs of the country. For this purpose the governments of the host countries should avoid introducing or maintaining inequalities of treatment between multinational enterprises and domestic enterprises in social matters affecting their respective workers.

91. Most Government members of industrialised market economy countries expressed the hope that such policies would help in taking full advantage of the positive aspects of the activities of multinational enterprises. In this spirit these Government members noted the recommendations of the ILO Tripartite Advisory Meeting on the Relationship of Multinational Enterprises and Social Policy, held in Geneva from 4 to 13 May 1976, that appropriate arrangements be made with a view to preparing an ILO Tripartite Declaration of Principles concerning Multinational Enterprises and Social Policy, which would provide the ILO's input into the much broader Code of Conduct which is currently considered by the United Nations Commission on Transnational Corporations. The interests of both the host countries and multinational enterprises were best served, in the long run, by an atmosphere of mutual trust, in which the rules for inter-relationship were known in advance and strictly observed, relevant information was available to all parties concerned, and negotiations were conducted in a flexible manner.

92. In the light of the above, the Government members of industrialised market economy countries were of the opinion that the present contributions of multinational enterprises to the creation of employment in the developing countries could be further increased through various measures such as:

(i)   local subcontracting when this was technically possible;

(ii) a progressive increase in the local processing of raw materials;

(iii) local reinvestment of profits to the greatest extent possible;

(iv) replacement of expatriates and maximum utilisation of local personnel;

(v)   training and promotion of local production workers and of local management personnel;

(vi) co-operation on matters of training between the multinational enterprises and the various local institutions providing training.

It should be understood, however, that the role the multinational enterprise could play in employment creation varied from one host country to another, from one time-period to another, and from one firm to another. On the other hand, the contribution of multinational corporations could only be partial since the reduction of unemployment in developing countries was a global task, the responsibility for which lay primarily with governments. It was therefore up to them to ensure that the contribution of multinational corporations to employment creation was maximised. The multinational enterprises should respect the sovereign rights of States as well as the relevant laws, rules and national practices and recognised international obligations, it being understood that it would be desirable to refer to Conventions and Recommendations of the ILO when legal, political and economic considerations so permitted. Multinational enterprises should adapt the activities of their subsidiaries to the development programmes and economic objectives of the countries where they were established. This adaptation should take into account all the economic and social factors of these countries.

93. Government members of the industrialised market economy countries considered that it was necessary to reinforce the technical negotiating capacity of developing countries vis-à-vis the multinational corporations. For this purpose:

(i) it recommended that the ILO should study regulations in the employment and training fields, adopted by developing countries, regarding foreign investment and multinational corporations;

(ii) it would be desirable to clarify the need for training in developing countries for the purpose of dealing with foreign investment and to establish corresponding training programmes which would assist governments in negotiating with multinationals on matters relating directly or indirectly to employment creation and the improvement of training;

(iii) it was desirable that the ILO, to the extent of its competence, should be ready to provide technical assistance as required in those fields to governments which requested it.

Also it would be desirable to ask the ILO to carry out studies on employment, training and wages policies adopted by developing countries regarding multinational enterprises. Research should equally be strengthened in the field of appropriate technology and labour-intensive goods, the production of which should be promoted in developing countries.

94. Certain Government members of developing countries associated themselves with most of the proposals in paragraphs 92 and 93 above.

95. Government members of industrialised market economy countries felt that multinational enterprises should so far as possible devote themselves to stepping up research and development in the field of appropriate technology and to the development of products to further employment creation. And that, lastly, for their part, governments should be able, before accepting the investment of multinationals on their territory, to be sure that the techniques proposed were those most suited to employment creation, taking account also of other factors affecting production and marketing.

96. Certain representatives of industrialised market economies, whilst in agreement with certain general points made in paragraphs 90 to 93 above, nevertheless expressed their sympathy vis-à-vis the declaration of the Group of 77. They also expressed their agreement with the procedures proposed in the Tripartite Advisory Meeting of May 1976, as well as with the proposal for research which the ILO could undertake in collaboration with the United Nations Commission on Transnational Corporations, without this implying, however, an acceptance of all the conclusions of that meeting. In addition they stated that it was necessary to co-ordinate the ILO's activities on multinational enterprises with those of the UN Commission on Transnational Corporations.

97. Certain governments, while recognising the importance of a Code of Conduct regulating the activities of multinational enterprises, put the stress on relations of a bilateral character which can exist between host countries and multinational enterprises and on the importance of national regulations for controlling the activities of these enterprises.

Declarations of the Employers' members

98. The Employers' members stated clearly that the relevant agenda item, as determined by the Governing Body at its 196th (May 1975) Session, called for a discussion of "the role of multinational enterprises in employment creation in the developing countries" and that they were prepared to discuss this specific question. They considered that companies in general, including multinational enterprises, as well as governments and trade unions, had a responsibility to bring about a better balance in the distribution of the world's products and knowledge. Multinational enterprises in conjunction with home and host governments and trade unions had an important role to play in advancing social progress. It was not possible for multinational enterprises to solve the problem of employment and to meet the basic needs of the world, but they had a contribution to make in this field; nevertheless, the responsibility of this task lay primarily with governments.

99. The Employers' members stressed that the discussion of the problem should concentrate on which kind of employment opportunities multinational enterprises could create. These enterprises did concern themselves with developing new activities important for employment, for example in agriculture. Although direct creation of employment by multinational enterprises was limited, the indirect effects were significant and could stimulate national economic development and know-how.

100. They believed that it was up to each government to decide what kinds of industrial activities and technologies were best suited to meet the development needs of its country. New activities of multinational enterprises

in developing countries should fit into national plans. Agriculture should be given priority attention in developing countries, and multinational enterprises could provide assistance in developing the production of industrial inputs to agriculture and in building up industries processing agricultural outputs.

101. The Employers' members stressed that multinational enterprises were a significant vehicle for the transfer of advanced technology, that choice of technology was often dictated by governments and that governments of developing countries generally insisted upon the most sophisticated kinds of technology.

102. They further expressed the view that multinational enterprises had beneficial effects on wages and working conditions. It was for host governments to define the social obligations under which multinational enterprises should function. It was the general practice of multinational enterprises to recognise workers' rights as well as the maintenance of labour standards and working conditions. In general, multinationals were responsible, did train local staff, had good industrial relations, had pay scales as good as, or better than, those of national companies, and worked within national regulations. Multinational enterprises were entitled to a fair remuneration for their efforts.

103. The Employer's members pointed out that multinational enterprises were free not to invest and that foreign investors needed a stable investment climate. Tough rules were acceptable as long as they were not arbitrarily changed. Moreover, multinational enterprises objected to regulations which were not applicable also to national companies. The Employers' members insisted on equal treatment on social matters.

104. Taking cognisance of the five reports prepared by the ILO at the request of the Tripartite Meeting on the Relationship between the Multinational Corporations and Social Policy which met in Geneva from 26 October to 4 November 1972 and of the agreed conclusions reached at the Tripartite Advisory Meeting on the Relationships of Multinational Enterprises and Social Policy of 4-12 May 1976, the Employers' members believed that it was not the mission of the World Employment Conference to discuss the content of principles to govern multinational enterprises. A voluntary code of conduct could be helpful.

105. The Employers' members considered that the ILO study on international principles and guidelines was a clear and comprehensive survey of possibilities in the ILO context. The ILO studies had shown that the multinational in general behaved responsibly. They had failed to reveal the existence of problems of the kind referred to by the Workers' members. The multinationals had been shown in the ILO studies to be a force for economic development. Indeed, they were the most effective means yet found for reducing the time-span for producing the management skills needed to organise resources

and muster finance. It was necessary to be careful that any action taken would not have adverse implications for the future. The Employers' members were therefore unconvinced of the need for international action in regard to multinationals in the social field. In particular, they considered that any move towards the adoption of an international labour Convention in this area risked creating an impossible situation through the variations in the extent of ratification or acceptance in different countries—a risk mentioned in the ILO study. There was also a question of discriminatory treatment. The bulk of the existing Conventions were of general application, the exceptions to this being so narrow in scope that there was no analogy between them and the wide range of enterprises and industries covered by the term "multinational", with their varying degrees of foreign and national ownership. A Convention applying to all employees of any enterprise under any degree of foreign ownership would place these employees under special regulations that might well be more favourable than those in the prevailing industrial economy of the country, with adverse effects on the orderly conduct of industrial relations. Having regard to the variety of industrial relations patterns and behaviour in different countries, the Employers' members believed that such matters must primarily be determined by the governments of the country concerned and the ordinary law and practice of the country.

106. Another approach that had been suggested was the preparation of a tripartite declaration of principles which could eventually be embodied in more comprehensive United Nations guidelines. The Office study had pointed to the guidance given in Conference resolutions and conclusions of Industrial Committees and other advisory meetings as indicating the feasibility of such a procedure. The Employers' members were not against guidelines in principle, as shown by those published by the International Chamber of Commerce as long ago as 1972 and the active participation of their organisation in OECD's work on a code. The Employers' members were, however, convinced that such a declaration would not be useful and might well be harmful unless the guidelines met the following points:

(a) that they ensure that the operations of multinational enterprises can continue effectively to the benefit of society as a whole;

(b) that they are non-mandatory but mutually agreed through a tripartite declaration of principles on responsible behaviour for multinational enterprises, governments and trade unions;

(c) that they ensure in social matters that all parties respect the laws and regulations of the host country;

(d) that they recognise the principle of equal treatment for foreign-owned and for national enterprises in matters of industrial relations and social policy;

(e) that they do not bind multinationals to observance of ILO standards not ratified or accepted by the host country, or introduce a system of standards

making existing ILO Conventions and Recommendations applicable only to multinational enterprises;

(f) that they are flexible enough to permit application to very different national situations and national objectives and in regard to widely different types of companies and industries;

(g) that they apply effectively to enterprises with public or mixed ownership as well as to privately owned companies.

Restrictive legislation would only slow down employment creation in developing countries by multinational enterprises. Multinational enterprises were already subject to many regulations and governments had adequate powers of their own, any of which could frustrate a company's expectations of a reasonable return.

107. The Employers' members stated that, following the proposal in paragraph 106 above, the Tripartite Advisory Meeting had recommended that a small tripartite group should be established to draft a voluntary declaration of principles applying to multinational enterprises, governments and trade unions. In view of this, the Employers' members did not consider it appropriate to place the question of multinational enterprises and social policy on the agenda of the International Labour Conference in 1978.

108. The Employers' members, after two weeks of discussion, were reluctantly forced to accept that no consensus existed in the group because the differing views of Government, Workers' and Employers' members were irreconcilable.

109. The representatives of employers of European socialist countries fully supported the point of view of the Government members of the European socialist countries with regard to the role of multinational enterprises in employment creation in developing countries.

Declarations of the Workers' members

110. The Workers' members expressed the concerns and preoccupations of trade unions and workers with regard to the effects of the activities of multinational enterprises on employment and more generally on development. They declared that the questions raised under item 4 in Chapter 11 of the Director-General's Report were not exhaustive and therefore should not limit the discussion. Consequently, the discussion ought to include other questions which were just as important. The Workers also underlined the fact that consideration of the problem should not be restricted by the conclusions of the Tripartite Advisory Meeting held in May 1976. Under these circumstances, the three international trade union federations asked that, on the international and national levels, steps should be taken to strengthen control of multinational

enterprises. This control should be exerted by the countries in which they operated. The areas in which international and national action should take place were, in particular, as follows:

(i) in all the countries where multinational enterprises operated, the existing Conventions of the ILO ought to be applied, in particular Conventions Nos. 87 on trade union liberties, 98 on collective bargaining, 100 on equal remuneration, 122 on employment, 135 on representation of workers, 140 on paid education leave and 143 on migrant workers. In addition, reference to ILO Conventions must include working conditions for multinational enterprises in countries which had not yet ratified these ILO standards and in those countries where they were persistently violated;

(ii) employment of local workers and non-discrimination should be guaranteed. Non-discriminatory working conditions should be established on a democratic basis and should correspond to the highest wages, salaries, working conditions and standards of hygiene and safety in all the branches and units of multinational enterprises;

(iii) multinational enterprises ought to guarantee that the enterprises supply the representatives of the workers with essential information, especially on the composition of capital, the general organisation of the company at the level of the parent company and the branches, the evolution of the company with respect to workers' participation, detailed investment plans, current and former agreements, conditions of work, wages and recruitment of personnel in each factory, data on financial management and results, etc.;

(iv) in addition, the right of trade unions to take solidarity action at the level of each factory and of the multinational organisation as a whole, and the right of trade unions to decide freely on any action designed to enforce economic sanctions;

(v) the transfer of activities following labour conflicts should be prohibited. In the case of a transfer of production, workers should be provided with new jobs with equivalent working conditions, and a compensation fund should be created to support workers losing their jobs;

(vi) furthermore, in a more general economic context, the profits of multinational enterprises should remain in the countries in which these enterprises operated in order to contribute to the creation of productive employment and to a healthier balance of payments situation.

111. The Workers' members felt that in order to achieve this, several convergent paths should be followed at both national and international levels. On the one hand, it would be desirable to strengthen legislative and executive powers to provide the possibility of prohibiting certain economic concentrations, to integrate the activities of the companies in national planning and to

provide for real public control over exchange, prices, monetary movements, investments, taxation and credit. On the other hand, the sovereign rights of States to nationalise in order to control their development and their sovereignty over natural resources should be respected. The right to nationalise should apply particularly when the interests of the workers or the country were threatened. Finally, it was necessary that a code of conduct should be elaborated at the international level defining the obligations of multinational enterprises. This code should take into account notably the principles and measures presented by the Workers' members. It should have a legal and binding form.

112. The Workers' members recognised the importance of the principle of non-discrimination between multinational enterprises and national companies in industrialised countries, but stressed that the very nature of multinational companies and the problems relating to them necessitated the possibility of making exceptions to this principle. In developing countries it was permissible and in some cases even necessary, in the interest of the development of these countries, to take measures which were discriminatory.

113. All foreign investments should be undertaken under the general conditions set out in paragraphs 110-112 and 114. In this context the multinational corporations should abide by the following principles:

(i)   local subcontracting when this is technically possible;

(ii) a progressive increase in the local processing of raw materials;

(iii) local reinvestment of profits to the greatest extent possible;

(iv) replacement of expatriates and maximum utilisation of local personnel;

(v)   training and promotion of local production workers and of local management personnel;

(vi) co-operation on matters of training with the various local institutions providing training.

114. Multinational enterprises should be required to study the manner in which they could adapt the activities of their subsidiaries to the development programmes and economic objectives of the countries where they were established. The multinational enterprises must respect the sovereign rights of States and take into consideration the legislation, regulations and relevant national practices as well as internationally recognised obligations. They must also recognise the rights of workers and should not undermine but contribute to progress in the field of standards and conditions of work in the host country.

115. As to future action of the ILO, a majority of the Workers' members insisted on the need to strengthen the technical capacity of developing countries to negotiate with multinational enterprises. In this field it was desirable that the ILO, to the extent of its competence, should be ready to provide the required

technical assistance to governments desiring to strengthen their bargaining power vis-à-vis multinational enterprises.

116. A large number of the Workers' members thought that it would also be desirable to request the ILO to carry out studies on policies concerning employment, training and wages followed by developing countries in relation to multinational enterprises. It would also be desirable to step up research in the field of appropriate technology and on products with a high employment content, the production of which it would be desirable to promote in the developing countries. For their part the multinational enterprises, so far as possible, should devote themselves to stepping up research and development in the field of appropriate technology and the development of products for furthering employment creation.

117. The Workers' members stressed that the ILO should deal with all the areas relating to the social aspects of the activities of multinational enterprises. The work of the ILO in these fields should be closely co-ordinated with the activities of the UN Commission on Transnational Corporations.

118. The Workers' members finally considered that:

(i) the ILO should continue its current work concerning multinationals and social policy on the basis of the conclusions of the Tripartite Advisory Meeting of 4-12 May 1976, but without confining itself to those conclusions;

(ii) the ILO should contribute in the field of its competence and within the United Nations to the elaboration of an international instrument (Code of Conduct) with a binding character permitting the control of multinational companies;

(iii) the ILO, within the framework of a reform of the mechanisms for examining questions concerning the violation of trade union freedom, should provide for a procedure to be applied to multinational corporations;

(iv) the ILO Governing Body should at its next meeting give consideration to the respective positions of the governments, the Employers' group and the Workers' group at the World Employment Conference;

(v) the ILO Governing Body should place the issue of multinational enterprises and social policy on the agenda of the 1978 Session of the International Labour Conference, in order that Conventions on multinational enterprises should be adopted in the following areas: industrial relations, training for employment, conditions of life and work.

119. The Workers' members expressed their profound dissatisfaction that it was not possible to reach any common points of agreement on this crucially important subject. They moreover wished to point out in this context that a

number of individual points of agreement were recorded between the Workers' members and several governments. The Workers' members expressed their support for the proposals of the Group of 77, in particular the basic principles covered by paragraph 85. They also supported points (i)—(vi) in paragraph 92 as proposed by the Government members of industrialised market economy countries.

INSTITUTIONS DEALING WITH APPROPRIATE TECHNOLOGY[1]

B

DEVELOPING COUNTRIES

Africa

Botswana

■ Appropriate Technology Centre

The purpose of the project is to create the means whereby knowledge about technologies appropriate to the needs and resources of Botswana can be identified, investigated, developed and applied.

In pursuance of the long-range and immediate objectives of the project and as a result of detailed discussions with government departments, parastatals and non-government organisations, four major problem areas (and, within these areas, a number of specific topics for early investigation) have been identified:

(a) *housing:* alternative forms of cement, small-scale production of Portland cement, the design of and materials for low-cost roofing, alternative materials for ceilings and partitions, low-cost door and window frames;

(b) *water:* low-cost supply to small rural communities, methods of low-cost small-scale storage, small-scale sewerage plants;

(c) *power:* examination of possible improvements to present methods of using fuel, utilisation of local coal as substitute for wood in rural areas, uses of solar energy (e.g. for water heating and pumping), uses of wind power (e.g. for water pumping), uses of methane (e.g. for water pumping and heating);

(d) *agro-industries and manufacturing industries:* small-scale processing of abattoir by-products (e.g. glue and brushes); grain storage in rural communities; small-scale manufacture of agricultural implements; small-scale exploitation of mineral deposits (e.g. cyanite, fluorspar); production of ornamental stone; small-scale foundry; small-scale enamelling; recovery of waste engine oil.

---

[1] Only institutions dealing mainly with small-scale and rural technologies appear in this list.

Technologies for basic needs

Many of the above topics have already been the subject of extensive R and D in various parts of the world, but as yet the appropriate technologies that have been developed have not been transferred or adapted to meet the needs of Botswana. The Centre will be a clearing-house and co-ordinator, an advisory unit and an evaluator, but at all times it will be concerned with the needs of users and not with technological research for its own sake.

It is felt that it would not be advisable or necessary for the Centre itself to carry out R and D work on the adaptation of technology, as a number of institutions equipped with adequate facilities to do so already exist in Botswana. The Centre would in particular encourage such institutions to solve local technological problems and provide them with the financial means to do so.

Ghana

■ Technology Consultancy Centre, Kumasi University of Science and Technology

The Centre was established in 1972 to serve as an intermediary between specialists in the University and potential users, and has been particularly concerned with small-scale industries. The Centre participates in R and D work by providing technical know-how and it assists in the testing of new products in pilot plants. It also provides technical assistance to firms in matters of quality control, commercial production, access to credit and the improvement of equipment.

The Centre has acquired a reputation for stimulating grass-roots development through the application of intermediate technology. Examples of such work include the upgrading of existing craft industries such as textiles, woodworking and pottery. The development of appropriate processes includes the manufacture of spider glue from cassava starch and alkali from plantation peel (these raw materials are abundant in Ghana) and the manufacture of broadlooms for village weavers. In the case of the manufacture of glue, the Centre provided technical know-how, a production plant and a financial loan to the entrepreneurs. In addition, the Centre has established three production units on the University campus for the manufacture of nuts and bolts, soap bars and broadlooms. The Centre's largest single project is a soap pilot plant. The Centre is engaged in commissioning seven small-scale soap-making plants (200-500 bars per day) using raw materials mainly of local origin and serving rural markets.

A programme for the establishment of craft centres in some 40 Ashanti villages was recently launched. Other rural non-farm activities include glass bead-making, the manufacture of coconut products, brass casting, and the local manufacture of such agricultural equipment as pumps, driers and bullock carts.

Kenya, Uganda, Tanzania

▪ East African Industrial Research Organisation (EAIRO)[1]

The Organisation covers Kenya, Uganda and Tanzania and is mainly oriented towards small-scale production units in the primary and secondary sectors. Food programmes in agriculture receive special attention through R and D work on the promotion of appropriate cultivation in different geographical areas (for example, the substitution of sorghum and millet for maize in semi-arid regions). Technical innovations are also carried out in industry (brickmaking, food processing, energy).

The work of the EAIRO is not biased in favour of the large-scale organised sectors, as is often the case with national and regional research institutions. Instead, it is in line with the basic-needs approach. Many of the smaller research projects are initiated in response to demands from small-scale clients.

Specific examples of innovations include:

(a) the development of a solar water-heating system which can be manufactured locally by small-scale sheet metal enterprises and which could provide hot water for domestic use in rural areas;

(b) the reconditioning of disused ceramics kilns for the manufacture of bricks and tiles;

(c) the development of techniques for the commercial manufacture of *oriatiô*, a natural dyestuff used in some dairy products; and

(d) the development of techniques for using Kisii stoneware for electrical insulators.

Nigeria

▪ Federal Institute of Industrial Research (FIIR)

The FIIR was established in the early 1950s to promote industrial development through local technical innovation for appropriate technology. Its main activities have been concerned with the modernisation of the technology for processing *gari* (a food item), which involved both quality improvement and large-scale production. Fundamental research was undertaken to study the processing of the basic ingredient and an appropriate system which resulted in the establishment of a pilot plant was subsequently developed. Further developments led to the full utilisation of the new *gari*-processing method by a farming co-operative.

In lieu of a proposed mechanised pilot plant, a completely hand-operated process superior to the existing traditional village method was developed. The mechanical grater designed and developed at the Institute soon became

---

[1] See ILO: *Employment, incomes and equality . . .*, op. cit., technical paper 9.

popular with the rural *gari*-producing communities, and its manufacture by rural carpenters and blacksmiths is now widespread. Another contribution by the Institute was the development of multiple frying range for *gari*.

Tanzania

- Small Industries Development Organisation (SIDO)

The Organisation was created in 1973 to replace the earlier National Small Industries Corporation. It is entrusted with the development and diffusion, on a regional basis, of appropriate technology for small-scale and rural industries.

The projects under way include food processing, building materials and clay products, textiles and clothing, and rural mechanised workshops. SIDO works closely with the ITDG, London, which is undertaking a project on bricks and tiles and another on the development of an appropriate leather and footwear industry on behalf of SIDO. Links are also maintained with the Government of India, which has provided assistance in the formulation of a hire-purchase scheme for small-scale industries. Furthermore, small-scale (OPS) sugar units are being imported from India.

Zambia

- Technology Development Advisory Unit (TDAU), University of Zambia

TDAU was created in 1975. Rural development, small-scale industries and low-cost housing are the three major areas of its activity. In the agricultural sector the Unit intends to concentrate on equipment for cultivation, harvesting and processing. It has so far developed two types of machine: a brick machine (for soil and sand/cement mixing) which is being introduced in rural areas through the Ministry of Agriculture, and a machine for extracting oil from cashew nuts. Future plans include the investigation of such subjects as the manufacture of paper glue using cassava starch and potash from husks of plantain, simple household electrical components and other household implements, nuts, bolts, and rivets, processes for soap production by adapting traditional methods, and the design of low-cost housing units and components.

Because of its present status as a unit within a university, TDAU seems to play a rather limited role in the application of appropriate technology. However, the recognition already received for its initial work is likely to permit its expansion into a full-fledged body able to: *(a)* collect information on low-cost, efficient machinery and processes; *(b)* develop new equipment either by adapting imported machines or by upgrading traditional ones; *(c)* conduct an evaluation of low-cost technologies in terms of market potential and cost competitiveness; and *(d)* ensure that specific needs for simple technologies are properly identified, through adequate links with appropriate ministries as well as entrepreneurs.

Asia

Bangladesh

▪ Appropriate Agricultural Technology Cell

This Cell was established in 1975 under the administration of the Bangladesh Agricultural Research Council. The objectives of the Cell cover the development and promotion of labour-intensive and capital-saving machinery and tools and implements for agricultural production; the manufacture of implements through the greater utilisation of local resources; and the development of appropriate drying, storage, processing and milling facilities to prevent post-harvest losses.

The principal activities of the Cell are the collection and dissemination of information on appropriate technologies for the rural sector and the initiation and promotion of research on rural technologies through grants to researchers in universities and other institutions. Working groups are established in the following fields: draught power; irrigation; fertiliser use and agronomical methods; post-harvest operations; and agricultural workshops.

Although this Cell is at present small and has a limited number of projects, it is proposed that its activities should be expanded into such fields as animal husbandry, village-based industries, rural housing, and so on. A proposal for the establishment of an autonomous Institute of Appropriate Agricultural Technologies is also under consideration.

India

▪ Appropriate Technology Cell, Ministry of Industrial Development

Set up in 1971, the Cell is composed of different working groups dealing with specific activities. Their objectives are:

(a) to assess the possibilities for scaling down plant size with respect to cement, sugar and paper;

(b) to identify technologies in road construction;

(c) to recommend a standard design and measures for dissemination of cow-dung gas (bio-gas) plants; and

(d) to identify appropriate tools and methods in building construction.

The working groups carry out mainly technological information surveys and distribute their results to potential users. They are composed of specialists from the country's leading research institutes. For a given industry or economic activity, all the production techniques in use are reviewed by these technical specialists. [1]

The Cell works in close collaboration with other organisations such as the

---

[1] See Government of India, Ministry of Industrial Development: *Appropriate technology for balanced regional development*, 2 vols. (New Delhi, 1975).

Council for Scientific and Industrial Research (CSIR), the National Research Development Council (NRDC) and the Committee for Science and Technology.

The apex body of the Cell is the Informal Advisory Committee, which includes members from a wide spectrum of interests, including industry, research institutes, government departments, specialised laboratories and the Planning Commission. This Committee meets quarterly to give direction to the work programme of the Cell and its related institutions.

● Programme on the Application of Science and Technology to Rural Areas (ASTRA), Indian Institute of Science, Bangalore

ASTRA was created in 1974 within the Indian Institute of Science in order to "serve as the agency for increasing the Institute's awareness of the rural situation; and play a key role in correcting the present urban bias in the educational, research and development programmes of the Institute so that a significant fraction of these programmes acquire a rural orientation".[1] The ASTRA programme is concerned with the development and promotion of appropriate technology for the satisfaction of basic needs, defined in terms of access to inexpensive essential goods and services for the unemployed and underemployed rural poor.

The first phase of the programme includes the development and testing of village-oriented technologies on the Institute campus; the establishment of an Extension Centre in a village near Bangalore; and the transfer of developed and tested technologies either to the village through the Extension Centre or to other rural development agencies.

The approach adopted is not to derive appropriate technologies for the rural poor by merely simplifying the modern techniques used in urban areas but rather, through the observation and study of rural traditional techniques, to improve them and increase their efficiency. The initial phase of the ASTRA programme thus involves extensive field surveys in order to identify the most crucial problems of the rural poor and the technical solutions to be envisaged.

Since 1974 a total of 12 projects have been launched, on windmills, handpumps, bullock carts, bicycles, rural housing, low-cost teaching materials, bio-gas, small-scale lime pozzolana, cement plants, the extraction of sodium silicate from rice husks, solar air-conditioning and Humphrey pumps. With regard to windmills, a specific design has been selected by ASTRA for its compatibility with the average wind pattern in India, as well as its appropriateness in terms of costs and possibility of local manufacture. Investigations are being made in order to improve and optimise the traditional type of cart. They concern the determination of the most appropriate type of wheel (wooden, rubber-tyred or pneumatic-tyred) and

---

[1] For details, see A. K. N. Reddy: *Problems in the generation of appropriate technologies,* Part II: *An Indian experiment,* Paper presented to the International Economic Association Conference on Economic Choice of Technologies in Developing Countries, Teheran, 18-23 Sep. 1976 (mimeographed), p. 5.

some changes in design which might be necessary if a new type of wheel were adopted. However, any modification of the design should meet a number of requirements, including acceptable manufacturing and maintenance costs, safety and stability under a wide range of operating conditions, ease of operation, and acceptable working conditions for the bullocks with efficient use of animal power. Concerning the supply of energy to villages, a number of studies have been carried out to determine the optimum size of plants using energy sources other than electricity, oil and gas. The result is that ASTRA is at present designing a community-size bio-gas plant for a village of 500-750 people.

- National Research Development Corporation (NRDC)[1]

The NRDC was created in 1953 with the objective of promoting the development and exploitation of indigenous inventions, processes and know-how from all sources within the country, with a view to achieving self-reliance.

The Corporation acts as an intermediary agent between public sector research organisations where most new processes and patents for development originate, and the enterprises which are potential users of these inventions. Recently the NRDC has engaged directly, and in co-operation with the inventor, in the manufacture of prototype production units and pilot plants (eight by 1974), in an attempt to overcome the reluctance of enterprises to adopt new, commercially unproven techniques.

The criteria for project selection include the project's potential in terms of export promotion, import substitution, exploitation of untapped indigenous resources, establishment of new industries, job creation, increased food production, and so on.

By 1974 the NRDC had conveyed 751 of its 1,175 registered inventions to about 1,000 enterprises; most of these (80 per cent) related to small- and medium-scale technologies.

The NRDC provides entrepreneurs with information on new processes through publications and technical journals, in particular *NRDC Processes,* a quarterly journal, which circulates information on processes for licensing.

Close liaison with industry and R and D institutions is maintained through participation in seminars, symposia and technical meetings.

- Planning Research and Action Institute (PRAI), Lucknow

The Institute was set up in 1954-55 with a grant from the Rockefeller Foundation. Its aim is to carry out research on the improvement and development of techniques for rural areas, in production activities as well as in "home living technologies". It deals with product selection, technology and organisational patterns. Activities include the design of pilot plants implementing new appropriate technologies for a given selected product. The

---

[1] See National Research Development Corporation of India: *NRDC Processes* (New Delhi), Jan. 1975.

Institute operates in co-operation with national bodies that specialise in the development of the corresponding product.

The diffusion of technology is also effected through seminars, publications and other measures (for example, technical and operational devices, training programmes for potential entrepreneurs and skilled workers); assistance is provided through the standardisation of machinery and equipment as well as the supervision of the manufacture of equipment.

The PRAI is also concerned with home living technologies designed to provide and improve basic services for the rural population, such as the construction of houses and the provision of water and fuel supplies, drainage and lighting.

In 1957 the Institute initiated a pilot project to scale down manufacturing processes. The result is a complete package plant which manufactures almost the same quality of sugar, with about 80 per cent of the efficiency of a large-scale mill in terms of recovery. The process, which has proved to be both technically and economically viable, has been applied to about 900 units in India. Similarly, a package plant for the manufacture of white-ware pottery for villages has been developed. A pilot project on evolving building products based on local raw materials is also in progress at present.[1]

■ Central Leather Research Institute (CLRI), Madras

The Institute was set up in 1953 and deals with all aspects of the leather industry, from the collection and preservation of raw materials to product and equipment design, technical improvement and the dissemination of information. It carries out R and D work as well as technical assistance activities through its main centre in Madras and its five regional extension centres.

Particular emphasis is given to activities at the small-scale and cottage level of the leather industry, and programmes are aimed at replacing exports of raw materials by exports of finished products.

R and D is carried out in the following areas: adapted preservation methods for new hides and skins; the improvement of existing leather in respect of shrinkage and colour fastness; tanning and finishing techniques as well as tanning materials; control of environmental pollution; and product design and development (leather garments, bags, cases, shoes and leather sports goods).

Technical assistance is provided through training courses on design development, sample making and fabrication; consultancy; the standardisation of processes and equipment; the preparation of feasibility reports for the setting up of manufacturing units; quality control; and the testing and evaluation of technical information.

The establishment of five regional centres has made it possible for a

---

[1] M. K. Garg: "Problems of developing appropriate technologies in India", in *Appropriate Technology* (London), Vol. 1, No. 1.

larger number of tanners to be instructed in the appropriate technique as elaborated by the Institute.

The Institute is also engaged in training university research staff. It publishes a monthly journal and various documents dealing with subjects related to the leather industry.

■ Appropriate Technology Development Unit, Gandhian Institute of Studies, Varanasi

The main aim of the Unit is to develop and disseminate information on technologies relevant to urban and rural poverty groups. It therefore engages in the following activities: agricultural tools and implements; food processing; handling and transport of materials; water and irrigation; decentralised sources of power and energy; construction; animal husbandry, and so on.

The Unit also promotes the introduction of appropriate technology in the primary and secondary school curricula. In order to encourage the universities to undertake R and D work on appropriate technology, it is proposed to establish a University Liaison Unit.

The primary aim of the Unit is to assist in raising the technological levels of cottage, village and small-scale industries and rural activities in general. In this task it acts as a catalyst in inducing the reorientation of the work of such agencies as government departments, industrial corporations and the Planning Commission.

One of the first publications of the Unit will be a directory of appropriate technology. This will list tools, equipment, plant and processes which are already being successfully used in India. The directory is intended to serve as a useful guide for field workers.

■ National Small Industries Corporation: Small Industry Service Institutes: Small Industry Extension Training Institute (SIET)

These organisations illustrate the Indian attempt to promote labour-intensive technologies through the establishment of small-scale industries. They provide technical assistance to firms as regards product identification and market surveys, training for small industry personnel, arrangements for the supply of raw materials and equipment, access to credit, and sales supervision and export promotion.

In all, some 100,000 firms have received assistance. The same number of people have been trained, and machinery to the value of US$80 million has been supplied on hire purchase.

Until 1960 the main activities of these organisations were concerned with the assistance given to entrepreneurs to ensure their access to material inputs and services. This assistance took the form of market studies and quality testing.

Since 1960 the SIET[1] has promoted training as well as research and servicing activities for small industry.

Training programmes include courses on the effective financing of small-scale industries, financial management, cost accounting systems, management accounting, materials management, quantitative techniques and reliability analyses, feasibility surveys and analyses as well as in-plant training. The SIET has also developed a course for unemployed engineers on product analysis and infrastructure for small-scale industries, in order to convert them into techno-entrepreneurs. International courses are also given on specialised items such as appropriate technology, capital and productivity, and materials management.

Examples of research work include the assessment of industrial potential in certain regions; feasibility reports on food items such as tamarind seed starch and maize starch; case studies on the evaluation of industrial co-operatives; a study of an Artisan Development Programme, including field studies of 300 artisans and a dozen trades; studies of the oil engine industry in Kohalpur and of the machine tool industry in Punjab; and a report on the hire purchase scheme of the National Small Industries Corporation. Finally, a Documentation Centre has been set up to disseminate information on different aspects of small industries to extension agents, entrepreneurs and other interested parties.

The Centre also publishes a quarterly *Newsletter* and a quarterly journal, *SIET Studies,* as well as monographs on specific subjects.

Indonesia

▪ R and D institutes (various ministries)

There are a number of research institutes, financed, operated and controlled by government ministries. Several of them, especially those under the Ministry of Industry, deal with specific industries, such as chemicals, ceramics, leather, batik and handicrafts.

Other ministries have similar research institutes—for example, fish processing, forest products, building materials and rubber.

One of the noteworthy features of these ministry institutes is that they are specifically industry-oriented and undertake only research which is of value to the industry served.

▪ Development Technology Centre, Institute of Technology, Bandung

The Centre's activities cover the planning, selection and development of appropriate technologies and of the specific skills necessary for integrated development. The principal areas of emphasis are as follows:

---

[1] See Small Industry Extension Training Institute: *10 years of SIET, 1962-1972* (Hyderabad, 1972).

(*a*) rural appropriate development technology. This programme proposes the establishment of a technical information system and a system of field stations for testing and demonstrating hardware or software technologies. Sample projects in hardware cover an agricultural product dryer, food processing (for example, coconut processing) and a stone cutter for the cottage jewellery industry;

(*b*) development of local power sources, for example, solar energy, wind energy, bio-gas;

(*c*) the regionalisation of technology transfer through regional development technology centres;

(*d*) a study, in collaboration with the Council for Asian Manpower Studies (CAMS), on the relationship between local small industries and multi-national joint ventures, emphasising vehicle assembly; and

(*e*) low cost housing—product development and testing of wood-based components for housing, surveys on housing markets and construction methods.

▪ Village Technology Unit (BUTSI)

BUTSI is a network of several thousand village-level workers. It publishes simple brochures in the Indonesian language on simple village technologies, such as solar stoves, treadle grinders for removing coconut meat, the making of twine from coconut fibres, and reforestation for flood control.

Pakistan

▪ Appropriate Technology Development Organisation

Established in 1973, the Organisation is designed to promote the concept and utilisation of appropriate technologies, which are defined as low cost, labour-intensive technologies intended to be more productive and based on the use of local materials. The Organisation is in charge of the collection and exchange of local and world information on appropriate intermediate technologies, the need and utilisation of which it assesses through surveys. The latter include a study of the availability of local inputs as well as the identification of potentially appropriate technologies in all sectors of the economy to be tested through pilot plants or field trials. When a technology has proved to be technically and economically feasible, the Organisation arranges for the design and manufacture of the plant and equipment required. It is also involved in training activities through educational and extension programmes for skilled and intermediate-level workers associated with the new appropriate technique.

The Organisation is primarily concerned with rural areas in order to encourage their development, through the promotion of employment. Priority topics have been identified as energy, water, food and agriculture, building construction and industrial chemicals. Projects now under way include the preparation of paper pulp from bananas, the development of windmills for raising water, low-cost housing, and the production of fertiliser and bio-gas from cow-dung.

Philippines

■ Regional Adaptive Technology Centre (RATC), Mindanao State University

The Centre is engaged mainly on work on appropriate technologies for small- and medium-scale industries, through R and D, training, consultation surveys and the establishment of pilot plants. It is divided into two sections: the first deals with the formulation of training programmes, conducts surveys and maintains contact with the community and local agencies; the second undertakes R and D and carries out the training programmes for specific industries.

The RATC's activities include the operation of a training-cum-production ceramics centre, a brassware centre, low-cost housing, fisheries and coconut charcoaling. The last three projects illustrate the development of indigenous innovations promoted by the RATC and disseminated in the country. They are based on the full utilisation of local resources applied to energy sources, food and housing.

Efforts are also being made by the RATC to introduce courses on small-scale industrialisation, entrepreneurship and related questions in the college curricula.

Sri Lanka

■ Appropriate Technology Group, Colombo

This Group is organised on the lines of the ITDG, London: that is, it is a coalition of Government, industry and universities with the broad objective of upgrading rural technologies. It was Dr. E. F. Schumacher's visit to Sri Lanka in 1975, sponsored by the ILO within the framework of the Asian Regional Team for Employment Promotion (ARTEP), that led to the creation of the Group.

The work programme of the Group at present covers the collection of information from firms specialising in the manufacture of engineering tools, the identification of specialised welding techniques, and the manufacture of dies and moulds for small-scale rubber mouldings, and of machine tools for woodworking. The Group's work is at present concentrated on the light engineering industry.

Thailand

■ Asian Institute of Technology (AIT), Bangkok

The AIT has engaged in a number of R and D projects, including a study of the mechanical properties of bamboo-reinforced slabs, the design of asbestos cement houses for low-income families, prefabrication in low-cost housing, the mechanical properties of wood cement composites and the design of low-cost rural school buildings. Other studies of a more theoretical nature have been carried out, including design and evaluation criteria for low-cost housing and the evaluation of a low-income multi-family housing system.

Latin America

Argentina

- Instituto Nacional de Tecnologia Industrial (INTI)

The INTI was established as a component of the new policy aiming to stimulate the development of indigenous technology, and is composed of a group of central laboratories in Buenos Aires and 20 research centres in the country. These are operated jointly with various industrial associations. The areas of activity covered include chemistry, physics, metallurgy, building materials, rubber, structures, food and textiles. The units also engage in product testing and analysis, but operate at the demand of the industry. The INTI provides management advice and direction, and recommends particular actions. Its activities are undertaken in close collaboration with the industries concerned, which contribute to the operating costs. However, the INTI is a predominantly consultative body and is essentially confined to research activities.

Colombia

- Instituto de Investigaciones Tecnológicas (IIT)

The Institute carries out R and D work for agriculture as well as industry. Its activities cover a large number of sectors, among which food processing has received particular attention. Product testing of food items to be launched on the market is undertaken in pilot plants; these pilot plants are also concerned with the improvement or innovation of processes and products. For example, the IIT has developed a portable sawmill for use in the tropical forests of Colombia where the transport of round wood necessitates the development of indigenous technology. Other examples of recent designs developed at the Institute include agricultural equipment, potato silos, food-processing equipment and building materials.[1] The Institute recently completed a study of the technical performance of the burned-clay products industry and has provided technological recommendations for ensuring an increase in the efficiency of small- and medium-scale brick producers.

---

[1] For more details on these see Instituto de Investigaciones Tecnológicas, Bogotá: "Capacity of the engineering industry in Colombia", op. cit.

DEVELOPED COUNTRIES

Canada

▪ Brace Research Institute (BRI)

BRI is essentially concerned with the questions related to energy sources in rural areas. R and D work has been undertaken into the development of equipment for desalination using natural energy sources (wind, sun); the practical implications of sun and wind for personal services (heating water and air) and agriculture (drying crops, as well as mills for water pumping and irrigation); the rationalisation of the use of natural energy sources. Recently the Institute began technological research on agricultural implements and processing industries.

BRI has also made information available through published surveys, and in 1974 its *Handbook on appropriate technology* was published; this *(a)* discusses the principle of appropriate technology and its criteria; *(b)* presents a number of case studies (solar distillation, integrated waste-fuel-food cycle, development of bee-keeping, small bio-gas plants, oil-drum cupola-foundry, and solar coffee dryers); *(c)* establishes a list of tools available to meet a wide variety of needs; *(d)* discusses future possibilities for new and experimental tools or systems currently under consideration for appropriate technology.

Netherlands

▪ Tool Foundation

This is a non-profit-making organisation which makes available to developing countries technical know-how and technology adapted to their local conditions. It is composed of a number of technical associated groups from the University of Technology, Eindhoven, the Appropriate Technology Group and the University of Technology, Delft.

The Foundation is concerned mainly with socially appropriate technologies. Small-scale agriculture and industry, alternative sources of energy, and simple tools and production methods are some of the priority areas.

The Foundation also runs an industrial inquiry service and a documentation centre.

United Kingdom

▪ Intermediate Technology Development Group (ITDG)

The Group's activities are carried out through panels and working groups dealing with specific subjects such as food, water, construction, building materials, chemicals, forestry, transport and rural health. These subjects are studied under their different aspects (administrative, industrial and commercial) and in collaboration with various specialised institutions. R

and D work, from major innovations to simple modifications, has been undertaken for a small-scale paper-pulp manufacturing unit, brick works, agricultural equipment, tanks and pumps. The results of this research have been published in the *Journal of Appropriate Technology.*

The dissemination of information is also undertaken by the ITDG through other channels, among which publication is particularly important; but field projects, consultancy and the creation of overseas intermediate technology units have also contributed to removing communication gaps in appropriate technology.

The ITDG has also established an Industrial Liaison Unit which tries to adapt industrial technologies available in the United Kingdom for small-scale projects in developing countries. In the latter, the Unit conducts feasibility studies and field trials for the local manufacture of small-scale equipment. Under a three-year grant from the Ministry of Overseas Development the research programme of the Unit was divided into a home-based unit which concentrated on liaison with industry in the United Kingdom and on answering inquiries from industrial users in the United Kingdom. The overseas unit was located in Nigeria, to identify opportunities for the local manufacture of small-scale equipment and to undertake research into the economic and technical feasibility of such equipment. For example, a feasibility study for hospital equipment has been prepared by Inter-Technology Services Ltd., a subsidiary of the ITDG.

An example of R and D work by the Unit is the redesigning of equipment used for weighing babies in the rural areas of East Africa. The Unit, in combination with Salter (manufacturers of weighing equipment), modified spring scales and designed plastic trousers and straps.

United States

• Volunteers in Technical Assistance (VITA)

This is an association of about 6,000 volunteer businessmen, scientists and engineers engaged in technology transfer in response to requests from developing countries. The volunteers have been helping in the design of simple bridges, the development of water supply and sanitation systems, seed selection and animal husbandry, and the promotion of small businesses.

A number of plans relating to the development of appropriate technology (for example, an inexpensive solar cooker and a simple well-drilling rig) have been compiled into the *Village technology handbook,* which is available in several languages.

VITA also runs an inquiry service and acts as an information collection centre. At present, it handles inquiries from more than 110 countries. Most of the inquiries for information relate to the small entrepreneur and to such subjects as housing and construction, alternative sources of energy, and medicine and health.

VITA has also established relations with a number of national organisations, such as the Village Technology Innovation Experiment (Ethiopia), the Industrial Development Board (Sri Lanka) and the Industrial Research Institute (Sudan).

■ Technology and Development Institute, East-West Center, Hawaii

This Institute is more of a research institution than a technical assistance organ, although it has institutional links with national institutions in the Pacific countries. In the light of advice from scholars and practitioners from Asia, the Pacific area and the United States, the Institute undertakes research and institutional co-operative activities relating to four main themes: *(a)* employment-oriented development planning; *(b)* the adaptation of technology; *(c)* the development of small-scale entrepreneurship; and *(d)* the development of public policy and institutions in the field of science and technology.

Research in progress includes work on low-cost construction (building materials and designs for housing), the development of the light engineering industry in Asia (with special reference to Indonesia, Malaysia, the Philippines and Thailand), and trade, technology and employment. Research work is undertaken by the Institute staff in close collaboration with national institutions. For example, in the project on low-cost building technologies, co-operation is maintained with the Institute of Science and Technology (Republic of Korea), the Asian Institute of Technology (Bangkok, Thailand) and the Institute of Technology (Bandung, Indonesia).

An effort is also being made by the Institute to identify and assess the feasibility of establishing regional adaptive technology centres. In this connection a research seminar was held in April 1974 to discuss and evaluate the findings and recommendations of field surveys conducted by the co-operating institutions.